# REAL STRANGER TRUTHS

SHANE D SMEDLEY

# DEDICATION

For Daniel
I have lived one life, died many times.

# CONTENTS

# ACKNOWLEDGMENTS

This book was made possible by men who, despite the danger involved, has made it his mission to share everything they learned while working in Special Access Programs. Without them and others like them we would still be living in the dark. As the old saying goes "truth is stranger than fiction". In this case, not only is it stranger, it truly can be frightening.

# THE NATURE OF TRUTH
# AND HOW TO MANIPULATE IT

### What IS Truth?

"Truth" is one of those peculiar concepts that cannot be precisely defined. The typical dictionary will state, "conformity with fact or reality." So you look up "fact" and find out that it is "something that is true" and end up running in circles. And we all know that "reality" is just an illusion, anyway.

"Truth" derives from the Latin *veritas*, a concept that means "verifiable," something that *actually exists* as *experience* or *observation*. What people miss is that *opinion* often masquerades as truth. So, how do these two get mixed up?

We obtain information on the world about us from two sources: our *sensations* (the spatial senses) and our *intuition* (the temporal senses). Sensation is valued by *thinking* and intuition is valued by *feeling*. (This is a classic Jungian concept of rational and irrational valuing.) Valuing, itself, is just how important we determine something to be, typically based on *ego gratification*. We give the highest value to things that make us feel good and give us power. So when it comes down to that juicy, fatty bronto burger or a Tofu latte… I know what I will choose!

Because all the information we receive from space and time are weighed and valued by *personal* needs and experience, we can never see the truth—so *truth is always subjective*, based on our extensive system of values, termed a *Weltanschauung* or *Worldview*. If you believe truth is objective, consider that mankind cannot even decide on the shape of the world—flat, round, flat again, oblate spheroid, geoid, hypersphere, n-dimensional membrane—whatever happens to be in vogue this year.

The Worldview is like a gigantic puzzle inside our minds with pieces missing and a bunch of them in the wrong place, because we don't really know what connects them to other sections. We've hooked together some bits of sky, that pond and that tree over in the field, because it was obvious those puzzle

pieces belonged together. But the holes between what we've assembled… that is what we seek to fill by searching for "truth." And we will only accept the "truth pieces" that actually fit in to the dents and wiggles of the pieces we've already assembled.

Many times, we encounter puzzle pieces that don't seem to fit anywhere, so we reject them as "falsehoods," even though they may not be. In this "disinformation age," the pieces we've already assembled were handed to us, pre-assembled, by those that wish to control what we seek when searching for the truth. Sometimes, we have to rip out whole sections of our puzzle because we just tried to place a skyscraper into a seascape, and no matter how we try, we'll never find the connecting pieces because they do not exist. But that never stops people from looking, and more importantly, *inventing* pieces to fit.

## Layers of Truth

There is another way we can fill in our gap-ridden worldview by stacking puzzles one on top of another. That way, if we have a hole in ours, we might be able to see the "answer" deeper down at a different level. This onion-skin wrapping of values is well known in psychology and there are *five* primary levels:

1.       The *personal* worldview, which is built from the experiences and observations of this life and the carryover from past incarnations (karma, unfinished business, etc). Choices made from our personal worldview are considered to be *free will* choices.

2.       The *familial* worldview that we inherited from our family and friends, normally taught to us as an early age as the "acceptable behavior" we refer to as *morality*.

3.       The *cultural* worldview, the generational knowledge of the culture we were raised within that is primarily a spiritual or religious worldview, from which we pattern our *ethical* behavior.

4.       The *societal* worldview, a body of proper behavior that is dictated by those rule us, to create social interactions with specific outcomes. *Laws* for the body, *codes of conduct* for the mind.

5.       The *species* worldview, inherent in our biology and archetypal makeup that come with just being human, forming our *instinctual* behavior. When all else fails, "the buck stops here" and we *react* instinctively, rather than *act* based on what we've learned. The species worldview is the only complete worldview—no holes.

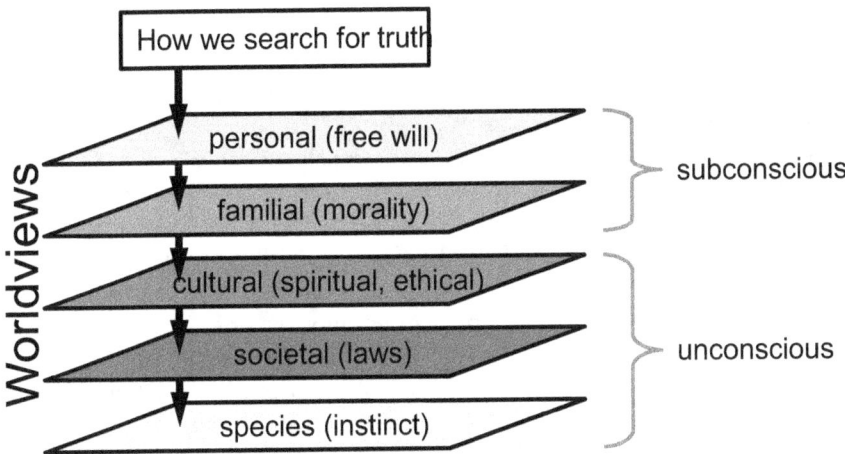

Prior to the 20th century, the *personal* worldview was extensive—one huge puzzle with many pieces filled in, as mankind was still curious about the world around him and engaged in extensive research, observing Nature and trying to duplicate what Nature was doing to understand how it worked. Researchers of that era had a difficult time obtaining and sharing information, as the communication technology we now have did not exist—no phones, Internet, television, radio talk shows, YouTube, etc. They were limited to a somewhat unreliable postal service, with international correspondence almost impossible—letters would take weeks or months to arrive, if they got there at all.

The 1950s brought the popularization of radio and television and the familial worldview began to fill in. Worldviews could now be shared with many people as stories. Families would sit together to listen to a radio play or watch *The Honeymooners* on the television—and tended to all watch the same shows, so behavior moved into the familial worldview and out of the personal. It wasn't because *you* wanted something this way, it was the way the family did it.

In the United States, a "melting pot" of culture, the cultural worldview never had much hold in the psyche because the United States never developed a *cultural identity* and immigrants were tending to flee the cultures they had lived in. In the mono-culture peoples, religion tended to be the controlling factor that is manipulated by those that run the churches—the first level of control of The Powers That Be.

The societal worldview is the most significant one, as mankind is a very social creature and needs a fixed set of rules for society interaction that crosses cultural and familial boundaries. And someone needs to write those rules—and those that made the rules became the rulers of society, originally the *Royalty* and now the *New World Order*.

The species worldview has remained fairly constant, but only provides the instinctual parts of the puzzle—how we behave when we do not have "learned behavior" to fall back upon. Because instinct is the basis of the survival mechanism, an inability to value at this level will normally result in death.

Now that we see how things "stack up," it becomes a fairly simple matter to

manipulate what fits into a
person's collective worldview—their "truth."

### What You Cannot Control, Destroy

The personal and familial worldviews are normally *subconscious*, that fuzzy area between active consciousness and the reactive unconscious. As such, they are difficult to modify because they are protected by the Ego. We all know how difficult it is to change someone's mind, particularly if they highly value the point in question. This also applies to the familial worldview, as the group ego will protect the morals learned by the family group (biologic or close friendship).

The species worldview is biologic, so it is very difficult to manipulate—unless you manipulate the underlying biology (genetics) of the species in question. Mankind has gotten to this point, but still lacks sufficient understanding of the system to effectively exploit it (lacks the 3D time concept of the mind).

The cultural and societal worldviews are behaviors that are *taught* by religious/spiritual and political institutions. As such, they are the easiest to manipulate through propaganda and marketing campaigns. But in order to reach them, one must get past the personal and familial worldviews—what you cannot control, you destroy. They must go, so *The Powers That Be* can get *The Power They Want*.

Religion and politics have gone hand-in-hand for centuries and not much can be done about the species, itself. So the first target up on the "hit list" was the *familial* worldview—the breakdown of the family unit and preventing close relationships from being formed between friends.

This started in the 1970s with the "village" movement, where it "takes a village" to raise a child—not parents. Add in daycare to pull the young, impressionable minds away so they only learn *societal* values and not *familial* ones. Anyone over 40 years of age has noticed what has happened to the family over the years and there is abundant material available on this topic, so details are not necessary. We just have to face that the conventional family concept, and associated worldview, are now gone.

That left the *personal* worldview, what used to be called *character*. Life experience taught us right from wrong, good from evil. In order to punch holes in the personal worldview, mankind needed to be disconnected from the experiences of life—Nature. And this was done through a concept known as over-socialization. Move any personal values into the societal worldview, by trickery, force or coercion, and *The Powers That Be* control those values.

The term "over-socialization" gets a lot of negative press, because if people became aware of what was going on, they would rebel against it and "make it personal." This is seen in concepts such as sovereignty and individuality movements, but is becoming increasingly rare because the tools used to suppress individuality are the ones misinterpreted as *rebelling against society*. And these are the same tools used in mind control: drugs (alcohol, tobacco, recreational, etc) and technology.

Drugs dull the mind to the point where valuing must rely on the unconscious for problem-solving, since familial valuing has already been destroyed. With enough use, the mind stops trying to store data in the personal worldview and defaults to societal programming—though the user will firmly *believe it is their free-will choice*. If you notice, there are very few "free thinkers" these days, but many thousands of "regurgitators" pushing that "share" button—without comment—in an attempt to simulate an impression of creative thought.

Technology has now become the "opiate of the masses." It has removed the biological need to remember information and experience, with the result of blanking out the personal worldview. It is rare for a person to actually have an opinion *of their own*, the most common response given is that "so and so said so." When it goes to "so and so," then you've fallen through a hole in the personal and familial worldviews, and dropped in to the programmable cultural and societal worldviews. "Truth" by consensus—not observation or experience. This is not *truth*, but a *deception*.

As I am fond of saying, *everything you know is backwards*. When you want to push a person's values away from something they would naturally be attracted to, you present it as its *opposite*, so that pull becomes a push. The societal worldview is deliberately designed to present everything backwards in order to keep you far from *veritas*, that which actually exists in Nature and can be observed and experienced.

**The Way Out**

The bulk of humanity is happy being *sheeple*, being told what to do and how to act. But if you're reading this paper, you fell through one of those rabbit holes and are seeking to find some answers to life, the Universe and everything. To do that, stop accepting what you are told is "truth," as it is nothing more than *opinion*. Engage in research as the old-timers did to find answers for yourself—and have *personal opinions*. And remember *veritas*—observation and experience leads to truth, which requires the use of your body, mind and spirit.

When you are confronted with an experience that does not fit into your worldviews, don't reject it because you can't figure out where that piece fits in the puzzle. Consider it… do a "what if" that piece actually belongs in this part of my worldview, then how does it change things? And if the change is better than what you already have, make it part of your personal worldview and see what happens. The Universe is actually a simple place based only on space and time, tending to follow Occam's Razor, "all other things being equal, the simplest solution is usually the right one." Truth makes understanding simpler, though on the surface, it may not seem so because one has to "undo" so many falsehoods that have been adopted as truth. As Dewey Larson, creator of the Reciprocal System of theory said, "Complexity is entertaining; simplicity is not."

Give simplicity a chance.

# THE UNCOMMITTED INVESTIGATOR

*Somewhere along the line, that which is true is being made to appear false, because that which is false is accepted as truth.*
—Dewey Larson

Larson's quote nicely sums up the current world situation, regardless of the field of study, whether it be science,[1] mathematics,[2] religion,[3] politics or the Great Pumpkin.[4]

One of the first questions people ask me after they find out what I know is, "why did you wait 30 years to say something?" Well, I *didn't!* For decades I tried to get people to listen, but no one would. I was insulted, attacked and ridiculed for daring to say that what people were taught in schools is flat-out *wrong.* Eventually, one concludes that it isn't worth the effort to try. So I stopped trying to get the information out and focused on understanding the things that I encountered during the Phoenix project, but as an "uncommitted investigator."

Dewey Larson,[5] one of my all-time favorite researchers and fellow "uncommitted investigator," said this in a presentation to a class of chemical engineers, back in

---

1 Dewey Larson's *Reciprocal System* demonstrate that conventional science tends to have things inside-out, upside-down and backwards, and when the situation is corrected, the Universe is a much simpler place than we thought it was.

2 The mathematics of Miles Mathes demonstrates the whole concept of infinitesimals is wrong; in nature, nothing ever "approaches zero," it *reaches* unity. Calculus became another dead-end, not an open door.

3 Research of forerunners like Lloyd Pye, Mauro Biglino and R.A. Boulay into anthropology show an entirely different history of mankind: a slave race engineered by the gods, not evolved from apes.

4 Schulz, Charles, *It's the Great Pumpkin, Charlie Brown*; From Linus Van Pelt's memorable line, "There are three things I have learned never to discuss with people: religion, politics and the Great Pumpkin."

5 Website: http://larson.rstheory.org/

1968:[6]

About twenty years ago Dr. James B. Conant, at that time president of Harvard University, gave a talk to a group of chemists and chemical executives in which he expressed serious concern over the effect on scientific progress that was likely to result from the virtual disappearance of what he called the "uncommitted investigators," a term which he applied to those individuals who carry on scientific research work on their own initiative, without support from or direction by the established research agencies. As Dr. Conant put it, these individuals "could investigate what they pleased when they pleased, or break off research at any point. They were as free as the wind because they had no program except the everchanging one in their own minds."

The reason for his concern, Dr. Conant explained, was that although the great majority of new discoveries in the scientific field are made by professional scientists working under the auspices of universities or research laboratories, the really revolutionary ideas, those that actually change the course of scientific progress, have come mainly from the free-wheeling activities of these *uncommitted investigators*, and if such individuals are no longer active, there is no assurance that these much-needed ideas will continue to materialize. In Dr. Conant's own words: "The revolutionary advances in theoretical science were made very largely by amateurs… Few will deny that it is relatively easy in science to fill in the details of a new area, once the frontier has been crossed. The crucial event is turning the unexpected corner. This is not given to most of us to do. If you want advances in the basic theories of chemistry and physics in the future comparable to those of the last two centuries, then it would seem essential that there continue to be people in a position to turn unexpected corners. By definition, the unexpected corner *cannot* be turned by any operation that is planned."

During my own research, I turned more than a few "unexpected corners" that pointed out, quite clearly, the things we are taught are "bits of truth," but *almost always lead to dead ends*. It is as though human knowledge has been guided into these dead ends and strong walls erected to keep people from thinking outside the box. True research, having been taken over by corporations, became a tool for profit—not a tool for understanding. It is only the few, uncommitted investigators working in back rooms, basements and garages, that have obtained new knowledge and tried to make it public—usually to be bought out by corporations, silenced under "national security," or ridiculed into obscurity.

The research I put forth is *not secret*, it is just the "common factor" to a lot of other research done by dozens of other uncommitted investigators over the last 200 years. The references I give can be found in most public libraries.

Take a dose of Larson, Maxwell, Steinmetz, Keely and Tesla, mix with some medieval Christianity,

Vedic epics and Hermes Trismegistus, shake well (shaken, not stirred), and season with a good dose of Sumerian mythology and Celtic folklore. What you get from

---

6 Larson, Dewey B., *Around Unexpected Corners*, 1968.

that concoction is the knowledge that was *forbidden* by the gods—the "secret science" that contradicts everything you know!

It takes the mind a while to unlearn everything it has been programmed to believe by those powers that be. My papers *directly contradict everything you've been taught* as "truth." For example:

- The substance of the universe is not *matter*, but *motion*, a abstract ratio of *change* that we call space/time.
- Faster-than-light speeds are *commonplace*; **the rule**, not the exception, which is a slap in the face to Einstein.
- Astronomers have everything *backwards*; stars begin their life as red giants and end it as blue giants exploding in supernovae.[7] Galaxies form from globular clusters, to irregulars, to spirals, to giant spherical structures and then explode, producing quasars.
- The geological dating system is based on a false premise; things are *much* younger than stated by geologists. Correlations are missed because of this, such as the Biblical appearance of Adam and Eve coinciding with the sudden appearance of Cro-Magnon man.
- Anthropology became the study of how to hide the true origins of *homo sapiens*. Fact is, mankind may have more in common with Godzilla than with the Neanderthals.
- Religion is not what you think… it's the code of enslavement, not enlightenment.

So when you read my papers and start screaming, "That's wrong! I was taught that…," remember that I've heard it all before. *I was taught it all before! And I used to believe it!* Then I found none of what I was taught came close to explaining the things I saw and worked with, forcing me to accept that *everything I knew was wrong.* It took a lot of time and effort to work around all those built-in biases— and they put up quite a fight—but once I was able to steer myself clear and take a clean look at what now seems obvious, a much simpler view of the universe emerged— one that opens the doors to some amazing possibilities.

The research contained in my papers is *not* channeled, received telepathically, supplied by ExtraTerrestrials, ExtraDimensionals, angels, demons, gods or any other source outside of a lot of *hard work* and good, old-fashioned "know-how," by a few dedicated, uncommitted investigators that want to *figure out* what is out there, not be handed a book on *The Universe for Dummies* by the next Vogon constructor ship that happens to pass by Earth, laying out a route for a hyperspace bypass.[8]

---

7 There are two types of supernova, based on the thermal limit (blue giant explosion) or the age limit (stars of other spectral class). See Larson's book, *Universe of Motion*, for details.

8 Adams, Douglas,
*The Hitchhikers*

This approach was a *matter of choice* by those involved, because we wanted to *demonstrate*, first hand, **to** those very ETs, EDs, angels, demons, gods, devils and whatever else is out there, that mankind *has learned to think for himself*, and is *ready, willing and able* to take his place in the Universe as a peaceful explorer that other worlds *want* as a good neighbor and friend.

I think we are all fed up with the violent, domineering agendas of the few reptilian wanna-be's that are out to control every aspect of our lives. It may be who THEY are, but it is *not* who WE are.

We are *Mankind*, the children of the Sons of God and the Daughters of Earth, unique in this galaxy. We *have a place* in the scheme of things, not as conquerers, not as slaves, but as *fellow travelers* on this Great Path that all life walks to understanding the mysteries of the Universe.

So Sayonara Saurians, *Homo sapiens* has had enough and is ascending without you, to stand as equals with those of the stars.

I started with a quote from Larson, and I'm going to finish with one, which is from the last paragraph in the book, *Universe of Motion*:

*The more complete understanding of physical existence opens the door to an exploration of existence as a whole, including those nonphysical areas that have hitherto had to be left to religion and related branches of thought. It is now evident that our familiar material world is not the whole of existence, as modern science would have us believe. It is only a part— perhaps a very small part—of a greater whole.*[9]

---

*Guide to the Galaxy* 9
      Larson, Dewey B., Universe of Motion, p. 438.

# GEOENGINEERING, CHEMTRAILS, HAARP, WORLD ORDERS AND ASCENSION

## Background

I was listening to exopolitics expert Alfred Lambremont on a radio show[9], discussing the "negative timeline" as being some kind of "synthetic quantum environment" orchestrated by the Grey ETs, stating that this negative timeline was all "made up." Having worked as a technician on the *Phoenix III* project (more commonly known as the *time travel* aspect of the *Montauk Project*), I have first-hand knowledge of what was going on and what they were trying to do—and it appears they are using some of the knowledge gleamed from those experiments in geoengineering application. In the 20 years that have passed since those experiments, I have learned much and decided to take a look at it in the context of the new knowledge provided by Dewey Larson's *Reciprocal System* of physical theory.[10]

*Phoenix III* was a covert attempt to determine the nature of time, if it could be manipulated, and if so, the process to do it. Without going into details,[11] the short answer is that those involved never came to an understanding of time, other than it was not what conventional physics thought it was, but were able to make use of some of the technology obtained in order to perform temporal experiments into the "past" and "future."

In order to access *time*, a navigation computer from a faster-than-light propulsion system was utilized to open a portal, from which a visual record could be made. In later experiments a way was found to send living organisms through this portal to other chronological periods, though this was highly dangerous and seldom

---

9 *Global Voice Radio* episode, "Are You on the Catastrophic or Positive Timeline?", August 13, 2012.

10 Reciprocal System of theory website (http://rstheory.org), was first published in the 1959 book, *The Structure of the Physical Universe*. The Reciprocal System is a *theory of everything*, based on two, simple postulates that produce the observed universe as a natural consequence. It has a number of notable successes, including predicting the existence of quasars some four years before they were actually discovered.

11 Details are covered in the paper, *Times and Timelines*.

worked as expected. Due to the nature of the equipment being used, only persons with a highly developed psionic ability could operate the equipment (psychics that were skilled in precognition.)

One of the issues was that the time travel experiments ran into a "bump," literally, crossing the 2012/2013 year boundary (conventional calendar) and at the time, it appeared to be some kind of artificial reality that was constructed by the psychic running the chair. The "experts" did not know what to make of it, but for the most part, thought it was a realistic view of the period being viewed because it was consistent with different psychics and everyone sent forward returned with very similar information. Looking at it now, it would seem to be more consistent with a "synthetic quantum environment," such as the one described by Lambremont, than a natural one, which may explain why it was always frozen past a specific point. With what we know now, there is a good deal of truth to what Lambremont was discussing on that radio show.

Research into the strange weather being seen around the world (including the effects thereof; red rain, abiotic stress of plants with "sudden death," extreme levels of radiation and ultraviolet light to name but a few) may have turned up what is going on... and it ties in with the old Montauk project. This interpretation explains the chemtrails, HAARP, geoengineering, underground cities, genetic engineering and even explains the timelines and the upcoming ascension to a higher density.[12] These connections cannot be seen by conventional science, due to various misconceptions treated as fact.[13]With the information in Dewey Larson's books, *Universe of Motion, Beyond Space and Time* and the follow-up articles by Prof. KVK Nehru on the nature of the stellar interior and the sunspots,[14] what is going on starts to make a lot of sense. You think the politicians hide stuff—try scientists!

Most of the things going on are a reaction to a larger event. Find the cause, and the effect starts to make sense. So, let's start with the most obvious effect: *climate*

---

12 We (the scientific underground), prefer *density* over *dimension* when describing the increased complexity of organisms, since the number of spatial (yang) dimensions does not change—the number of available temporal (yin) dimensions, does, resulting in a compaction that has properties similar to the physical concept of density.

13 Some of the misconceptions being: the backward direction of stellar evolution, the omission of 3-dimensional time, the omission of the "cosmic sector" (the universe of antimatter), and the nature of radiation and x-ray emission.

14 KVK, Nehru, "Glimpses into the Structure of the Sun" series; *Reciprocity* XVII, #2 (Autumn 1988), Part 1, "The Nature of Stellar Matter" and *Reciprocity* XVIII, #1 (Winter, 1988), Part 2, "The Solar Interior and the Sunspots."

*change*. And for that, we have to take a look at the single, largest effect on weather—sunlight—and where it comes from, the **sun**.

## Stellar Evolution is Backwards!

In *Reciprocal System* (RS) physics, Larson comes up with a better structure for the atom, based on the concept of *scalar motion*, and as a natural consequence of that structure determines two "destructive limits" for atoms: a *thermal limit* and an *age limit*. The thermal limit is the common limit used in thermo-nuclear reactions, but the age limit is *unknown* to conventional science and concerns the capture of charged, electron neutrinos by the atoms that create isotopic mass.[15] When the age limit is reached,[16] an element becomes radioactive and explodes. When Larson applied those concepts to astronomy, he found a *different* mechanism for stellar combustion based on those limits. Combustion is based on exploding atoms—*fission*, not fusion—and that led to the conclusion that astronomers have the stellar (and galactic) evolutionary sequences *backwards*!

In *Universe of Motion*,[17] Larson shows the consequences of what happens when you flip stellar evolution around, making red giants the *youngest* stars, and the blue giants the *oldest*.

Backwards evolution made for a perfectly logical and consistent pattern of evolution eliminating the need for a bunch of devices that conventional astronomy introduced to try to make sense of their interpretation; things like *dark matter, dark energy, black holes, quark stars, neutron stars…* when you put the sequence right, they all become just distinct stages in a single, consistent, stellar evolutionary process.

As a consequence to setting the stellar evolutionary sequence straight, some conclusions are different than popular belief:

- Stars start out as large dust clouds, condense into red supergiants, orange giants, yellow then white main sequence stars, then on up to blue giants, then supernova. The same process as heating up a piece of metal, from the initial red glow to where it gets blue-hot and breaks.

- Since stars are built from dust and debris as fuel, the more fuel available, the hotter the sun will get.

*Stellar Evolution is Backwards!*

- As David Wilcock has identified in the past, planetary moons are leaving "trails" throughout the solar system, indicating we've entered a

---

15 Larson, Dewey B., *Basic Properties of Matter*, Chapter 24, "Isotopes."

16 Larson calls it an "age limit" because atomic process is similar to aging. The capture of neutrinos is inevitable over time, and irreversible, so eventually the isotopic mass becomes greater than the structure of the atom can contain, and the atom "dies" by a temporal explosion that is viewed in space as radioactive emission.

17 Larson, Dewey B., *Universe of Motion*, Chapter 4, "The Giant Star Cycle."

dusty area of space. With all this dust and debris available for stellar fuel, the sun is *growing in size* and *getting hotter*, moving from a class-G (yellow) star to a class-F (yellow-white) star.

• Because our system of long-term dating is based on an assumption of radioactive elements being formed when the Earth was born—not being produced regularly—our system of geologic eras is drastically wrong. Astronomical events move thousands of times faster than assumed by astronomers. The planet and solar system are nowhere near as old as claimed and mankind has been around to see the sun "change" in the past.

When you consider these consequences from a corrected stellar evolution, one conclusion is obvious: global warming is not due to flatulent cows, but due to the fact the *sun is getting hotter*, and will continue to get hotter.[18]

**Sun Heats in Quantum Steps, not a Smooth Transition**

In the *Reciprocal System*, everything is quantized into discrete units. So is the case with the destructive limits... stars do not heat smoothly, but tend to stay at a specific temperature, then suddenly jump to a new temperature range as the magnetic ionization level increases (which controls the age limit). That is why we have a distinct color-temperature class system for identifying stars. The discrete jumps become very visible when you look at an H-R diagram with the correct evolutionary sequence, as stars move from red supergiants, to orange giants, to main sequence—distinct bars on the graph with few stars between them. The earlier stages of heating up take bigger jumps than the later ones, so that is obvious here.[19] Astronomers consider the giants to be separate from dwarfs because they do not realize the stars heat and condense in quantum steps—they assume a continuous change, so they miss the connection.

**Radioactivity: Accelerating to FTL Speeds**

Conventional understanding of radioactive elements needs some updating. In the RS, when you accelerate matter *past the speed of light* it becomes *radioactive*—it emits radio waves while it is throwing off particles. That's why they call it "radio active." It has to do with the fact that the zone of isotopic stability inverts when you cross the speed of light. Atomic explosions in the stellar interior are violent enough to push motion past the speed of light, something that *cannot* be accomplished by

---

18 About 9 minutes in to Jesse Ventura's *Conspiracy Theory* television program, "Global warming" (season 1, episode 3), he interviews a reclusive "climate scientist" who has come to the same conclusion—it's the *sun*, not man, that is the cause of warming.

19 Has to do with the way magnetic ionization affects elements as a 2nd power function. With no ionization, elements up to 118 are stable. With unit ionization, everything from *uranium* up becomes radioactive—27 elements, which is current ionization level for Earth. At 2 units, everything from *gold* on up is radioactive—only 13 more elements added to the fuel source, and it diminishes from there.

electromagnetic means in particle accelerators.

For example, take U-236. Uranium is element #92, so its natural mass is 184 (twice the element number) and the remaining 54 units of mass are "isotopic mass," an accumulation of mass (neutrinos) that forms the 236 isotope. 184+52=236. Once U-236 is accelerated past the speed of light, the inversion takes place and the stable zone becomes 184-52 = 132. The atom has to throw off 104 units of mass (2x52) to become stable at FTL speed. This throwing off of isotopic mass is *radioactive emission*.

If you think back to the days of Bob Lazar and his claims that UFO propulsion used element 115… well, 115 has a natural mass of 230 and an estimated mass of 288, meaning there are 58 units of isotopic mass. 115 is highly unstable at speeds less than light, but only has a mass of 172 at FTL speeds —*completely stable*. Pull it out of the reactor, though, and you'll be bombarded with x-rays to the point of glowing in the dark. This is why the FTL drives are *never* shut down. Which brings me to the reciprocal process—matter dropping from FTL to sub-light speeds.

**X-Ray Emission: FTL to Sub-light Speed**

When high-speed matter drops below the speed of light, it must reacquire the isotopic mass it lost. In U-236, the FTL mass was only 132. When it drops back to sub-light speed, that mass must *increase* back up to 236 amu, which is kind of backwards radio-active emission on the other side of the speed of light boundary—the atom absorbs particles and *emits x-rays*, not radio waves, as it builds mass back. All elements dropping from FTL motion to sub-light will emit x-rays, and all astronomical x-ray emitters are demonstrating this process—including our sun.

The only sub-light speeds in the sun are in the photosphere. Once you get deeper inside, the magnetic ionization level is much higher and the age limit destructions are constantly accelerating matter to FTL speeds, which is why the lower layers of the photosphere are a radio source—it is the boundary to FTL motion. Every now and again, the sun burps[20] and some FTL matter comes out from the core to the photosphere, immediately starts cooling and drops below the speed of light generating a burst of x-rays and a rapidly expanding plasma—a *coronal mass ejection*. Because of the reciprocal relation, FTL motion is *expanding* in time, so *compressing* in space. When it drops sub-light, that compression reexpands like a spatial explosion. So CMEs are a good indicator of how turbulent the core is at FTL velocities.

Now you know that the reciprocal process to radioactive emission is x-ray emission, and both have to do with crossing the FTL boundary (sub-light motion in 3D space to FTL motion in 3D time). FTL acceleration produces radioactivity; deceleration produces x-rays.

**The Solar Transition**

Time to put the pieces together… the sun is getting hotter from all the dust and

---

20 Described in detail in Prof. KVK Nehru's paper, "Glimpses into the Structure of the Sun." See footnote 28.

debris the solar system is now experiencing.[21] The increased fuel will increase the thermal destructive limit, which will cause a corresponding increase in magnetic ionization level, which will make more elements available for the stellar combustion process—the sun is going to get *brighter* and *hotter*. Initially, this will occur as bright flashes,[22] like a mini novae, until sufficient material is available to hold the magnetic ionization limit at the next quantum step. At that time, the sun will **suddenly** jump up in stellar class, and remain there. (Well, "up" in the Reciprocal System, "down" in conventional astronomy, since they have it backwards.)

*The Solar Transition*

The transition should be interesting. When the magnetic ionization level of the sun increases, it will be like throwing a cup of gasoline on the barbecue grill coals—a burst of flame and thermal activity, so much that it will move the thermal speeds past the speed of light. This "inverse thermal emission" actually occurs frequently on a small scale and is documented in detail in Prof. KVK Nehru's paper, *Glimpses Into the Structure of the Sun: The Solar Interior and the Sunspots*, and is the reason that sunspots are dark and appear cool. Inverse (FTL) thermal motion is super-hot, so hot that it *appears* cold and the region of the sun where it takes place goes dark, as in the sunspot umbra. There are already indications of this beginning to occur.[23] Except this time, the whole sun will become an "umbra"— there should be a bright flash, like a nova flare, when the gas hits the fire (additional elements suddenly being available for fuel from the jump in magnetic ionization), then the sun will *go dark*, like it went out. But only for a short time, until the initial burst of new fuel has burned up and the sun returns to the zone of stability. Like most things, this has happened before[24] and will happen again. Also recall the radioactive transitions. When the magnetic ionization increases,

---

21 The reason for the excess dust in the solar system is not clear, but astronomers have known it was coming for over 50 years. New Agers have billed it as the *photon belt*, but it is more likely to be just nebular dust or a protostar that is intersecting the plane of our solar system. As our sun, part of the Sagittarius Dwarf, intersects the plane of the Milky Way, such occurrences are very probable.

22 These bright flashes of the sun have been reported since 2010 at the higher altitudes, where the atmosphere is thinner and there is no pollution/smog layer. The increased intensity (flash) lasts for several minutes to several hours and tends to be laser-like, causing peculiar damage in a small area, such the sudden death of plants and trees (leaves are burned to a crisp) or the cracking of car windshields.

23 NASA spots giant, triangle-shaped dark spot on sun

24 Mythological references to "three days of darkness" may have their origin here.

there will be a huge burst of radio waves as the material is accelerated FTL, along with the nova flare. The sun will go dark —FTL motion—and when it starts to light up again, there will be a huge burst of x-rays from the sun, and the possible ejection of a great deal of matter from the surface of the sun, due to the re-expansion of FTL thermal motion back to sub-light speed.

**Post-Transition**

After the transition is complete, the sun will be physically larger, brighter (more white than yellow) and hotter than before—and it is going to stay that way. One would think that this situation would make the inner planets go up like marshmallows burning on the campfire. But curiously, that is not the case. Seems that whoever designed stars and planets considered this, and used the energy of transition to aid in the further evolution of life.

What will happen is that, due to the increased FTL motion in the sun, the gravitational balance of the solar system will change. FTL motion is anti-gravitational, so the sun will literally push the planets further outward in their orbits in compensation—the year will get longer.[25] Being further from the sun, the planet will survive and establish a new ecosystem—but a different one.

The changes in the sun will also produce changes in the planets, particularly the electro-magnetic alignment of the poles. As has been noted in geologic records, the north and south poles of the planets have been in various locations across the globe—not because the poles are moving, but because the crust of the planet is moving relative to the mantle and core. There is also a high degree of probability that the event will trigger a "core flare", an expansion event of FTL matter in the core dropping to sublight speeds and causing the crust of the Earth to expand and open at tectonic boundaries, eventuating in more surface area and a drop in ocean levels as compensation.[26]

In my opinion, this solar transition is the "harvest" or "ascension" to a new state for life on Earth—not just man—*all* life on Earth. All the physical properties get "kicked up a notch," as Emeril would say, commonly known as a "higher density" or a "higher dimension."

**A Hot Time in the Old World, Tonight**

If you were NASA, the *National Aeronautics and Space Administration*, or the "powers that be" and knew the sun was constantly getting hotter,[27] but did not know the specifics because you listened to conventional scientists with their backwards stellar evolution, what would you be likely to do? Obviously, come up with a long-

---

25 This has been recorded in mythology as occurring a number of times, from the 260 days, to 360, to our current 365 day year.

26 Peret, Bruce, "At the Earth's Core: The Geophysics of Planetary Evolution", *Reciprocity* XXVII, № 1 (Spring, 1998).

27 Larson gave many lectures to NASA back in the 1960s when they were trying to get into space—so highly probable they know what is going on; at least in the upper echelon.

term strategy to deal with the excess heat. Back in the 1950s, a study was made on such a possibility and came up with three alternatives:

1.        Use nuclear weapons to blow holes in the upper atmosphere to let the increasing heat out, 2.    Create huge, underground cities to live in until the sun stabilized, or

3.        Get the heck off of Earth.

The first one was a bit preposterous and could not be pulled off without public knowledge, so they started the 2nd and 3rd.

The underground bunker approach was simple enough, as one of the features of HAARP[28], the *High Frequency Active Auroral Research Program*, was that it was able to probe well beneath the surface of the earth looking for oil and minerals resources, but also identified large caverns that would make excellent cities for them to stockpile and ride out the transition in comfort.[29] Construction of these underground bunkers has been in progress for decades, and *Alternative 2* is nearing completion.

NASA came into existence shortly after this report was issued; some believe as a direct consequence of *Alternative 3*. Their mission was to find out what was "out there," and what they found was that mankind is not going anywhere else, any time soon.[30]

With *Alternative 3* not viable and knowing they would be stuck here with the rest of us, and not particularly wanting to live underground all their lives, they needed to come up with a way to make their Utopia on Earth… and that is *geoengineering*. Change this planet to be resistant to the solar changes they knew were coming and keep their corporatocracy[31] going.

**Geoengineering: Customizing the Planet**

First thing they had to deal with was the sun getting a lot brighter from the combustion of dust and debris, now present in significant quantities in the solar system. They needed to come up with a way to create a "global dimming" effect

---

28 After discovering that the ionosphere reacted strongly to HAARP transmissions, the project was militarized (see US Patent 4,686,605, "Method and apparatus for altering a region in the earth's atmosphere, ionosphere, and/or magnetosphere") and can now create ionospheric lenses that can be used to focus the sun's rays on a specific area, causing a localized rise in ground, ocean or atmospheric temperatures. This updraft creates a low pressure system at low altitudes and a high pressure system at high altitudes, and can be used to modify the flow of weather patterns.

29 Alex Jones of InfoWars.com reported these underground cities are large enough for 80,000 people with supplies for 30 years.

30 What they found "out there" will be addressed in a separate paper. Mankind is far from alone and not well received.

31 *Corporatocracy* is the rule by corporations.

to block off this bright light. Aluminum, a nice, lightweight and very abundant element, works rather well for that as most of our mirrors today are coated with aluminum (not silver). Nanoparticles, distributed in the tropopause (about 7 miles up[32]), would increase the albedo of the Earth and turn the upper atmosphere into a partially-reflecting mirror.[33] This has the result of

*Geoengineering: Customizing the Planet*

"global dimming" on the surface.

Next problem is the x-ray bursts. The Van Allen belts protect the Earth from most particulate radiation that comes from CMEs, but that magnetic field barely slows down x-rays. Fortunately, back in the 1950s a technology became available that had to address a similar problem—the *Cathode Ray Tube*, or CRT. CRTs were a large x-ray emitter, pointed right at the person watching the screen. To get this technology out, they had to develop something to block x-rays but not interfere with the picture. They tried lead, but after being bombarded with x-rays for a while, a lead film turned brown and messed things up. It was OK for the sides of the tube, but not the picture screen. Searching around, they found a couple of oxides that did an excellent job at blocking x-rays and stayed viable for a long time without browning out. That was a combination of *barium* and *strontium*.[34]

They also expect huge particulate radiation from both the CMEs and the solar heat-up, so they looked for a way to increase the magnetic flux in the upper atmosphere, which has been popularized now as "magnetic reconnection"—not a natural phenomenon. Aluminum, barium and strontium are paramagnetic and what was really needed was a ferromagnetic material... how about another popular metal, like *iron*? Might make the sky and rain turn a bit reddish on occasion from iron oxide, but you could just say the people who saw that were crazy. And so what if thunderstorms got massively more violent from the conductive metals in the atmosphere (super-cell storms). People don't pay attention to things like that, right?[35]

So they developed a technique to disburse aluminum, barium, strontium and iron

---

32 The aircraft observed dumping "chemtrails" are normally at 35,000-40,000 feet—6½ to 7½ miles.

33 Could not make it orbital, as solar radiation would vaporize aluminum and blow it away with the solar wind.

34 They use to list CRTs as highly toxic because of this—if you broke one you had to make sure not to touch the broken glass, as it was considered poisonous.

35 Abundant iron provides a generous growth medium for ferrous-feeding bacterias, particularly MAC (*mycobacterium avium complex*), which is the cause of many respiratory problems these days. Just a convenient side-effect for the pharmaceutical companies to profit on.

in the stratosphere using the 10,000+ aircraft[36] that are in the sky, every hour of every day. All they had to do was develop an aerosol and fuel additive, and "let 'er fly" to "git 'er-done." Couple the aerosol dispersion system with a GPS (Global Positioning System) and you can even control precisely where the chemicals get dumped, without the pilots ever realizing anything is happening.

Now if you look at chemtrail fallout in snow, ice and rain... what do you find? Aluminum hydroxide, barium oxide, strontium oxide and iron oxide. The same elements listed on climate modification and geoengineering patents. What a coincidence.

As to the question of "global warming"... yes, the planet *is* heating up. However, due to the global dimming[37] created by the chemtrail project, that heating up actually turned to a bit of cooling for a short time, causing conspiracy theorists to say the whole global warming thing was a farce. But if you've been outside lately, it is *obvious* that the sun is significantly brighter than it used to be, particularly at the higher elevations. And the chemtrail sunscreen is failing, though they constantly double efforts to reinforce it.

The experts are well aware of the sun becoming brighter and hotter, but you can't get the sun to pay "carbon taxes" for causing global warming. But if "man" is the source of the global warming, then you've got yourself one major cash cow with carbon taxes, environmental research, cleanup, breathing taxes, drinking taxes, waste reclamation taxes... to paraphrase Carl Sagan, "billions and billions" of bucks.

The initial geoengineering work was to block the solar changes and was just a temporary solution until the solar transition completed. But, like good humanitarians, the globalists saw a different picture... what if they destroyed the natural cycle and introduced an artificial one? One that they were the *sole supplier* for? And the only way you could survive was to pay through the nose.

**Geoengineering and GMOs**

The globalists have always profited on the pain and suffering of others, and this is no exception. They did not bother to put much research into the "side effects" of dumping millions of pounds of these elements in the atmosphere—they only considered the immediate need to protect from the solar transition and figured they could take care of the side effects later on. But here was a major opportunity in the form of disasters, pain and suffering.

A couple of the major side-effects were *drought* and *superstorms*. One cannot tip the seesaw, without having extremes.

Drought occurs in the areas that are being heavily seeded, as these nanoparticles in the upper atmosphere form condensation nuclei for rain—but there are *so many* nuclei, they never condense sufficient water on them to get heavy enough to fall

---

36 http://planefinder.net/, which tracks every plane with a transponder—the volume of air traffic is amazing, to say the least.

37 Global dimming on Wikipedia here:
http://en.wikipedia.org/wiki/Global_dimming for some propaganda on global dimming, but right now, about 20% of the sunlight is being blocked.

out of the sky, at least not right away. They just stay up there as water vapor, drifting with the air currents, creating drought conditions below. Given enough exposure to moisture over time, they will condense sufficiently to form rain. Due to the significantly higher volume of nuclei, will have a much higher "rain density" than a natural storm, so when it rains, it *pours*—the superstorm. This creates an unnatural dichotomy; upwind you will have drought, downwind you will have excessive rain.

Also, nanoparticle vapor could be knocked out of the sky with an application of RF energy to shake things up a bit and cause condensation and rain. Thermal vibration causes collisions, and collisions have the tendency to merge particles. Experiments along this nature has been conducted by Arco Power Technologies[30] over the years for weather modification that resulted in HAARP, which had all the tools to do this. These metals in the upper atmosphere made it incredibly simple to control the weather, planet-wise. If you control the weather, you control the world.

Controlling the world through weather has its own side-effects, particularly since it tends to poison all the life on the planet. In order to keep things going—and under their control—the genetic makeup of life had to be altered to be tolerant to the new, planetary climate—particularly aluminum and barium (natural strontium is actually beneficial—makes better bones than calcium).

One of the things that must be kept in mind is that this started in the 1950s. It has been in progress for some time and if you think all those cattle mutilations were done by ETs with anal probes, think again. Cows have blood and organs that are compatible with *homo sapiens*. Great for genetic experimentation. Once the basics were worked out, they moved to humans, trying to find a better genetic design for man to live in this new, artificially-created environment that will continue to exist after the solar transition. This gave rise to all the UFO stories about human-alien hybrids. It was all just part of the genetic

---

30 *APTI, Inc.*, the company that built HAARP, is assignee on patents, such as:
5,202,689: *Lightweight focusing reflector for space*,
5,041,834: *Artificial ionospheric mirror composed of a plasma layer which can be tilted*,
4,999,637: *Creation of artificial ionization clouds above the earth*, 4,873,928: *Nuclear-sized explosions without radiation.*
*Geoengineering and GMOs*

engineering that had to go hand-in-hand with the geoengineering.

If you look at all the weird UFO stuff going on over the last 60 years, most of it was nothing more than steps towards the focus of creating an artificial world and artificial population, to continue 3rd density life on the Earth after the solar transition.

And it was not just people. Long-term food sources had to be adjusted to remain viable in the artificial environment, and that's where your good friends at Monsanto came in with their drought-resistant, toxic chemical resistant, GMO foods. Sure, experiment with the mass population to find the good strains for the people who deserve them. And make sure you have HIPAA and "health care" to get all that genetic data back to the computers, to determine what worked and what didn't. After all, they're not collecting mountains of medical data, they are protecting your privacy!

They are really quite happy now with their little, artificial empire and are engaging in all the stalling tactics they can. The solar transition is already under way and all they have to do is keep the masses distracted long enough for the good stuff to start so they can lock themselves away in their underground bases and let the rest of us burn.

Or so they think. To quote Dr. Malcolm from *Jurassic Park*... "Life will find a way..."

## ELE and Ascension

*Extinction Level Events*, ELEs, are not an "end," but simply a time of change. Happens all the time in nature. Happened with dinosaurs, the hominids, uncountable millions of insects, bugs and varieties of plant life. Recent summaries say that the number of species going extinct has increased 1000-fold, compared to a century ago. But they aren't just disappearing... they are evolving. And that is what is supposed to be happening with mankind and the other intelligent life forms of Earth. (Humans are only the *dominant* species, not the only intelligent one.)

If the globalists had not started messing around with geoengineering and genetics some 50 years ago, humanity would have already been in the transition phase to a more complex form of life—a "higher density" expression of consciousness. All the GMO foodstuffs we are exposed to on a daily basis, combined with the chemicals used in climate modification, has delayed certain natural processes from being initiated.

In Dewey Larson's book, *Beyond Space and Time*, he discusses the "life unit," the origins of the living cell and the biological level of existence.[38] One of the more salient points concerning biological structure is that it *mimics stellar behavior*—life is composed of a stable combination of matter and antimatter (antimatter being called "cosmic matter" in the *Reciprocal System* that exists in 3dimensional time). This linkage also has its sub-light (spatial body) and faster-than-light components (temporal mind or soul), and even continues one step beyond that into the realm of ethics and metaphysics—hence Larson's title, *Beyond Space and Time*.

Life has remarkable similarities to the various stages of stellar evolution observed by astronomers, when the evolutionary direction is corrected. Stars are born from a cloud of dirt make a body, the infusion of a soul as the compression of matter reaches faster-than-light motion, and its eventual death in the glory of a supernova. Life can be thought of as twin suns, one in space and one in time, linked together in a controlled explosion of energy, such that all the thermal extremes cancel each other out. As Delenn from *Babylon 5* stated, "we are star stuff"[39] and are inexorably tied to the stellar and planetary processes of the solar system. We have scratched the surface of these associations with concepts such as astrology and metaphysics, but only scratched.

The problem that has arisen is that for the last 50 years or so we've been blocked

---

38 Larson, Dewey B., *Beyond Space and Time*, pp. 70-88.

39 A paraphrasing of a notable comment by Carl Sagan.

off from our genetic inheritance by geo- and genetic engineering. The proper signals to do a "life unit upgrade" were not received when they should have been, kind of like those people at the gym that wear headphones and don't see the basketball bouncing towards their face, despite people yelling to "look out!" Ker-smack, and they are totally caught by surprise.

Time to rip the headphones off of people, and let them hear the warning signs—and the signal to "upgrade" their psyche. There is still hope, but I don't mean from ETs...

## Resetting Genetic Modifications

I have been doing some of my own, anti-GMO experimenting to see what it would take to get a "terminator seed" to germinate, based on the concepts discussed with David Wilcock in 2005, regarding the Russian "torsion fields" experiments to modify DNA. And I have had some success. Genetically-modified organisms, when exposed to a torsion field broadcasting "heritage seed" DNA, literally throws out the artificial changes and reverts *back* to the original DNA coding and germinates. Only about 15% success at the moment, but the important point is that it *does work*. And the seeds produced continue to germinate on their own. This, alone, gives me hope that if the artificial conditions being imposed on the planet were removed, life would quickly revert back to its "factory defaults," pick up the proper signals, and begin the upgrade process—what is called "ascension." After all, the Earth's core is the largest torsion field generator in existence on this world—someone just needs to push the "reset" button.

Geoengineering *requires* genetic modification to keep the status quo. If removed, nature will reassert itself and start the process of healing. The altered atmosphere and genetically modified organism are seen by nature as an *injury* and *disease*—and it can, and *will*, treat those injuries and kill the disease when given a chance.

The globalists must keep inflicting injury to the ecosystem in order to retain control of it. If they don't, they'll lose control of it and it will start to revert to what it should be.

## Time Lines

About these negative and positive time lines that Lambremont and others are talking about... Having worked so much with the idea of a reciprocal relation between space and time, and it's Eastern counterpart, yin-yang, these negative and positive time lines need to be considered two *aspects* of the evolutionary process. It isn't going to be one or the other, it is going to be **both**.

The solar transition must occur, as it is a completely natural process that is observed all over the universe. It just needs to be understood that when the sun evolves, the life associated with it also evolves—the life on all its planets, moons, and everything inbetween.

What is being called the *negative* time line is the one the globalists want, their artificial environment with total control. If one remains 3rd density after the solar transition, that's exactly what they are going to have to have in order for 3rd density life to continue on Earth, as the conditions they are preparing for will actually exist—*but only in 3rd density*. And odds are, they won't last long, and the remaining 3rd density life on Earth will either die out, or have to be relocated to another world. (It is fear based, and

*Time Lines*

they could certainly use a shoot-out between the "good guys" and the banksters to get that going. People living in fear will be stuck in 3rd density.)

The positive time line is the one of ascension, moving to 4th density and beyond. The solar transition provides the energy needed to push life forward in its evolutionary process. Granted, with Larson's research I can identify the mechanics behind it, all the way through the biological stages. But that is actually unnecessary once the evolutive connection becomes evident. The sun evolves, which caused the planets to evolve, which causes life to evolve. Everybody wins. To a person going *with* the process, all they see is growth and improvement. To those left behind, they see chaos and destruction. You can be clobbered by a big wave on the ocean, or grab your surf board and have a great time riding it out. It is a personal choice.

The difficulty faced by the ecosystem of Earth is the unnatural tampering at a critical time. They may be able to mess up the atmosphere and genetics on the surface of the planet, but they still cannot touch the inner workings. And that is what people need to do—get back in touch with the planet, as a living, intelligent entity.

My working with plants and animals here have proved to me, beyond all reasonable doubt, that a reconnection to Earth initiates the transition and ascension process, regardless of what the globalists are doing. It puts us back on the positive growth line. I've literally resurrected burned-out trees from abiotic shock, which are now green and thriving in the new environment *without* genetic modification. In extreme drought and excessive heat, I still have squash plants about to invade the next county. No GMO needed, though a little "Miracle Grow" *was*, if you get the double entendre.

## Epilogue

Like almost every aspect in our society, it still comes down to a "natural" versus "artificial" condition, whether it be a natural person versus an artificial, "corporate" person, real money versus fiat currency, or a natural, evolutionary world versus an artificial, 3D slave society.

I know there is a lot of talk about divine intervention, ETs, angels and whatnot coming to "save us"… sit back, the cavalry is coming. Being of Cherokee descent, that never sounded very good to me. But all mankind really needs is an anti-globalist inoculation, so these monsters dry up and blow away. I won't call them human, because after what I've seen at Montauk, I don't believe they are.

Personally, I don't want ETs intervening in our society, dumping their version of "absolute truth" on us. One thing I've learned in my life is that nothing is "absolute"… everything is measured relative to something else. It cannot be absolute truth, just truth relative to what the ETs have accepted as truth.

What bothers a lot of people is that many of us have spent most of our lives looking for truth, and now some alien is going to drop by and hand us a copy of *The Universe for Dummies*, which will do nothing more than make us feel like our entire life was wasted. I don't mind a few pointers, but *I want to do the work for myself*.

Now if the ETs want to come by with a big vacuum and suck off all the parasites that have retarded human consciousness and evolution, I'll be the first to shake their hand, claw, tentacle or whatever. But I don't really want to get rid of a world order, just to have it replaced by a stellar order, reeducating and reprogramming me with their truths so I don't have to think for myself.

Consciousness doesn't grow from being *told* facts—it grows from the *search* for them. And I think that is what the people of this world really need—a chance to grow up on their own in a free society.

Hopefully, you can use this information to help people get that opportunity.

**Addendum: Hurricanes**

Using the information on the structure of the sun provided by Dewey Larson and Prof. KVK Nehru, one can infer a similar structure to planets (see footnote 18, page 5). Though the planets have an atmosphere rather than a photosphere, what goes on beneath the mantle is very similar to what goes on in the core of the sun—and the Earth produces similar effects as the sun, but just on a *different scale*. One of the more prominent features of the sun is that of the sunspots, dark areas that often produce flares and prominences on a recurring cycle. Recent footage has also shown that "solar tornadoes" are present above these active areas of the sun, stretching thousands of miles.

If the Earth were generating its own version of sunspots, what would we see? First, a regular cycle, but at a different scale, rather than every 11.5 years, perhaps once a year. And when a sunspot formed, there would be a massive, vorticular motion to it that would pull up material from underneath and send it sailing out around it, just as the solar tornado does: we just call it a *hurricane*.

If one examines Figure 8, "Migration of Prominences" of Nehru's paper on "The Solar Interior and the Sunspots," it shows that hurricanes show up at the same latitudes as sunspots, move in the same fashion, and diminish just as Nehru describes the sunspot cycle. There is a very high correlation. Hurricanes, like sunspots, are just a feature of planets. These are seen on other planets as well, such as the giant red spot on Jupiter—a hurricane.

Knowing that hurricanes are the product of "co-magnetic thredules" in the Earth's core, they are natural events and *cannot* be artificially created. Nor can they be directed in their early stages, when the magnetic forces are at their strongest. But they *can* be intensified during formation by the *same* process that creates the super-cell thunderstorm mentioned earlier—excessive amounts of water vapor nuclei forced into the upper atmosphere through chemical seeding of the tropopause to create vast quantities of clouds and torrential rain. When the generating thredule begins to collapse, the hurricane can be controlled by localized changes in temperature and pressure, just what HAARP does best.

This was observed in the case of hurricane Sandy. A small, tropical depression that never got to more than a Category 1 hurricane. Nothing spectacular about that, and common for this time of year—*except* the *quantity of water vapor present* in the system was significantly higher than it should have been. Watch the animations of the storm as it grows—the center of the storm is literally throwing off massive amounts of clouds because the small amount of ocean water being pulled up met the excessive amounts of condensation nuclei that were sent to that location via atmospheric chemicals—chemtrails.

The co-magnetic thredule creating this storm lasted longer than normal, which is why the storm followed the "sunspot" track, rather than being deflected by the Gulf Stream back out to sea, as hurricanes normally are. This indicates that, like the solar transition, the Earth is getting ready for a similar change down deep. HAARP would not have been able to direct or deflect this storm with an active thredule, though they probably tried. It followed a natural course, but an unnatural intensity, thanks to chemtrail geoengineering.

# TIME AND TIMELINES

### The Nature of *Time*

One of the least understood concepts known to man is that of *time*. A great deal of headway regarding the nature of time was made by engineer Dewey B. Larson, published in his 1959 book, *The Structure of the Physical Universe*.[40] Larson asserts that both space and time are simply the *aspects* of a reciprocal ratio that he refers to as *motion*,[41] have no other meaning, and cannot exist independently outside of this relation. He often drew an analogy to a box, with the outside being *space*, the inside being *time*, and the box being *motion*. If you have an inside and outside, then you have a box. If you have a box, then you have an inside and an outside. If you have an outside then you have an inside; an inside then an outside. So it is with space (outside), time (inside) and motion (box). The three concepts are always connected and cannot operate independently. Larson's theory eventually became known as the *Reciprocal System of physical theory*.

In the reevaluation of the Reciprocal System, RS2,[42] it is pointed out that the concept of space-time is analogous to the Eastern concepts of yin-yang, where space is the *yang* aspect, and time the *yin*. As discussed in taijitu symbolism, yin-yang cannot be separated, just as Larson's ratio of space to time cannot be separated.

---

40 *The Structure of the Physical Universe* is an incomplete work that is currently out of print, superseded by the 3-volume set, *Nothing But Motion*, *Basic Properties of Matter* and *Universe of Motion*.

41 Larson originally used "space-time," and later changed the label to "motion" as not to be confused with the common, coordinate representation of space-time used by conventional physics.

42 The Reevaluation of the Reciprocal System of theory, website: http://rs2theory.org

Larson then took it one step further, observing that all the characteristics of space must also have a similar character in time. In space, we see a 3-dimensional, coordinate spatial grid with clock time. There, from the reciprocal perspective, there must also exist a realm that contains *3-dimensional, coordinate time* with *clock space*. He refers to the former as the "material sector" and the latter as the "cosmic sector," identified in conventional physics as the *universe of antimatter*.[43]

The material sector is our common reference frame, that contains the observable and measurable structures of the universe. The cosmic sector, however, remains unobservable and unmeasurable to our physical senses, though we can see its effects on how *time changes space*, those effects being called *force fields* (electric and magnetic fields). We cannot see a magnetic line of force until it interacts with a material object such as iron filings, and alters their behavior in space.

So what we have in the *Reciprocal System* model of the universe is two different sectors of existence, the 3D spatial, material sector of our common experience, and a 3D temporal, cosmic sector that we cannot directly perceive, but is still there, influencing space.

With a proper understanding of time, the concept of timelines can be understood as a *path* through a 3dimensional, temporal landscape, where the ideas of past, present and future are just *abstractions* of that path in a temporal landscape: what is behind you is the past, where you are standing is the present, and what is in front of you is the future.

### Project Camelot Video: David Wilcock Interviews Bill Wood

As documented in the video, "Bill Wood: Live Q&A,"[44] Wood apparently came in on the tail end of everything, but I can clarify a few of the things discussed.

As Wood mentions, they [the *New World Order*] do not understand "time," specifically the distinction between "clock time" and "coordinate time." Three-dimensional time, like its spatial equivalent, has coordinates. In coordinate time, you have "clock space" (our concept of distance). Wilcock[45] mentions this about 45 minutes in when talking about moving into coordinate time, walking a distance, and translating back to coordinate space to appear in a different clock time. That is essentially correct and it gives some insight in to what the "clock" actually is—it is a "scaling factor" that our consciousness uses to "scale" coordinate time to appear as a "temporal distance"—*duration*. It is a similar mechanism to space,

---

43 In the *Reciprocal System*, it is technically "inverse matter", not antimatter, since the relation is the multiplicative inverse, not the additive inverse inferred by the prefix, "anti-".

.

44 Youtube, "Bill Wood: Live Q&A,"
http://www.youtube.com/watch?v=9k7J0RWLFGo

where doubling the distance to an object looks the same as halving its size (translation versus scaling). The NWO folks don't understand this and try to apply 1D temporal vectors in a 3D temporal system—that resulted in the necessity for a device referred to as a *Temporal Vector Generator*, or TVG.

About 58 minutes in, Wilcock states the purpose of the TVG was to align the timelines… not exactly, though it can affect them. The TVG was a device to *navigate* 3D time, just like you would target an object in space with a gun—except they target a temporal coordinate. Once you have the coordinates of where you are and where you want to go, you can plot a course from one to the other—the tunnel.[46] The TVG, itself, is not much more than a surveyor's instrument for the temporal landscape.

If you were to target a spatial location with a gun and shoot something, the spatial landscape is altered. So it is with sending something to a targeted coordinate *in 3D time*. It is what you *send* that alters the timeline… not the orientation of vectors. Again, as a spatial analogy, it is easy to blow up a building in the distance, but pretty hard to shoot an "undo bomb" to put it back the way it was. Same thing in the temporal landscape. If you blow something up there, you cannot put it back to the way it was because "past" and "future" are abstractions, not actualities. You are stuck with the alteration.

Structures in the temporal landscape are unaffected by what we consider "clock time"—there, it is "clock space" that acts in a manner analogous to the conventional clock.[47] Once they start messing around in the coordinate time landscape, they cannot undo what they did and have try to keep manipulating nearby regions to alter the flow to where they want things to go. It starts this oscillation, bouncing from one side to another, as Wood described. They are always overcompensating and have just about lost control and literally "destroyed" the future (the distant part of the temporal landscape they were targeting, *not causality*).

As mentioned, one must have the coordinates of where you are before you can plot a course to another

*Phoenix III, an Investigation into the Nature of Time*

temporal location. As discussed with Wilcock years ago, this was the ZTR, the *Zero Time Reference*. This reference was established for the *Phoenix III* project during the *Philadelphia experiment*, when they sent a ship back to 1943—the ZTR for the artificial realities is centered on 1943 and they have about a 50-year range of fire, so to speak. In time, you don't fire spatial structures—you fire "waves," because the coordinate time realm is *aetheric*—time appears solid, because it is the

---

46 Exactly like Daniel Jackson, in the original *Stargate* movie, described the 7th chevron as the "point of origin."

47 The material sector is based on the space/time relationship, how space changes with respect to time. In the cosmic sector, the situation is reversed: structure is based on the time/space relationship, how *time* changes with respect to *space*, hence "clock space."

reciprocal of the spatial, material sector, which is empty. Coordinate space is "empty" with locations filled by time; coordinate time is "full" with locations emptied by space.[48]

So their "guns" are like a couple of people playing a flute to a remote observer. Even though they may both play the note "C", the frequencies will not be *exactly* the same, so the further you go, the more the waves go out of phase. Far enough, they may even cancel each other out from that phase difference. This is where the concept of *range* comes from. The computer equipment they use to generate these waves is through digital-to-analog conversion. Even with very high resolution, it is still a digital reconstruction… there are errors that get worse, the further in time it goes.

### Through the Looking Glass

Now let's hop over to the *Looking Glass*. The *Orion Cube* is SM[49] technology. But *Looking Glass* is a reverse-engineering of the *Golden Sun Disc of Mu* that is mentioned in George Hunt-Williamson's books.[50] Don't know here the original device got to, but it was similar to the "ring transporter" on *Stargate SG-1*. (I think all the "portals" mentioned in the video are *Looking Glass* ring transporters— not interplanetary stargates. As far as I know, Earth only has a single gate address, and only *one* functional device, down in Antarctica.) The Sun Disc also had the ability to "remote view," which was how a destination was determined by a priest that would use it for travel. Using song, mantra and music, the priest could fine-tune the device to a very specific destination that could be seen in the device, before actually making the transit. (It works with coordinate time, hence controlled by waveforms—music, mantra, chanting, singing and the like).

The *OBIT*[51] "all-seeing eye" was an associated technology. They found that each

---

48 This concept of "empty space" and "solid time" is very important to understanding the reciprocal structure of the universe. Atoms are literally "solids of time" positioned in the coordinate vacuum of space. We then perceive the cosmic sector, the universe of antimatter, as "holes of space" in the coordinate solid of time, which is the origin of the aether theories of the 19th century.

49 "SM" is a designation to refer to the *Saurian Men* or *Space Men*, often called "reptilians." (Their actual species is not within the taxonomy classifications of terrestrial life.)

50 Brother Philip, *Secrets of the Andes*, edited by Timothy Beckley and Brent Raynes, 1976.

51 OBIT is an abbreviation for *Outer Band Individuated Teletracer*, which was leaked to the press by its designers due to ethical considerations. An *Outer Limits* episode was created with the same name for "plausible deniability," as the ethical concerns regarding the use of this device were staggering.

place has a kind of DNA resonance to it. In the old days, it was temples and monuments—the large, stone constructs made for a very stable resonance that was easy to find with the Sun Disc. The frequencies here were on the "inner band" of the inanimate realm. Once they started reverse engineering the system, they found they could tune up to the "outer band" of the biologic and pick up on specific DNA resonance of *any life form*, giving them the ability to locate anyone, anywhere.

The problems they have with the reverse engineered devices, both extraterrestrial and ancient, is that the devices engineered by man only contain *inanimate* structure—3D space, only. They have not yet realized that the "water" they have to use in these devices is *living water*[52]—all life has a presence in 3D time and creates a natural crossing between the sectors. The third component is the consciousness of the operator—they must *comprehend* coordinate time principles. The old priests created the discipline of *Hermetics* and the "river of time" analogy to describe this function of consciousness. When you try to view 3D time from a space-only perspective, you switch from a linear, *step measure* (equal intervals in a straight line) to a polar step measure (equal angles resulting in *growth measure*), which *appear* as an *infinite series*. Draw a line on a sheet of paper; put a dot above the line. Connect the dot to the line with radii… if you equally space the angles between radii, you'll see where the radii hit the line at *unequal* distances—they get longer and longer the further you get from the orthogonal, and eventually spread to infinity when parallel.

The reason they called it *Looking Glass*, aside from Alice in Wonderland's ability to transport to Wonderland, was that it was also a *hall of mirrors*. Because they assumed linear measurement in a polar, aetheric realm, they got reflections upon reflections upon reflections—each looking similar, yet slightly distorted by the observer. This gave the impression of "parallel universes." The distortions were introduced by the consciousness of the observer, but in actuality they were just viewing a distorted recursion of the coordinate realm.

I'll follow-up with some commentary on the timelines and how to alter them—and even how to get your butt out of the predefined ones.

## Timelines

The setback that occurs in many of these projects is the failure to understand the properties of **time**. And fortunately for us,[53] they are usually too arrogant to admit that their science is inadequate! Those in charge of these projects *always* listen to the "experts." And my definition of an "expert": *a person that knows more and more, about less and less, until he knows everything about nothing.*

---

52 See the research of Viktor Schauberger, and the book, *Living Water*, by Olaf Alexandersson.

53 "Us" meaning the *scientific underground*, those people who work to decode and understand ancient and extraterrestrial science and spirituality without the knowledge of governments or the general public.

As a result, the incorrect concept of time being *linear* and *vectorial* (the "arrow of time" stuff) predominates scientific thought—and is taught in all the schools, so most people never think beyond that box. In order to understand the "timelines," one must understand 3D time as a *temporal landscape*, and in that landscape, the "future" is what is in *front* of you, and the "past" is what is *behind* you. The "present" is where you are standing in the temporal landscape. And I would like to clarify that—the "present" **does not change**. If you were to freeze your presence in the moment called "now," you would be eternal. But it would also be rather boring, as *nothing would ever change*! So when considering the concept of past, present and future, try thinking of it as an orientation in a 3dimensional, temporal landscape—and your position is constantly shifting, ever so slightly, even when believe you are "still."

Consider the implications of this scenario: just like in space, no two people can exist at the same point in coordinate time; there is always some separation. This infers that each person's *view* of the past and future is slightly *different*... the general features of the terrain (mountains, rivers, valleys, etc) can be agreed upon, but things up close can have a radically different perspective. If two people turn to face each other, then they are destined to meet in the future (what is in front of them)—just a few temporal steps away.[54]

Now consider the psionic side (the psychic, metaphysical or ESP side). In space, we can see a car a half mile away, driving down a road heading towards us and consider it "normal." In time, if one sees a car

*Timelines*

heading towards us in the temporal landscape, we call it "precognition." Precognition is the ability to see a distance (technically a "duration") in the coordinate time landscape—Larson's concept of *clock space*. If you and I were standing 10 seconds apart in that temporal landscape, that car might hit you and miss me, even though we "predicted an impact," because we both saw it heading our way. We could discuss the approaching car, and people in the spatial realm would think we are telepathic— because we are standing right next to each other *in time*, and our spatial bodies could be miles apart. The "timelines" are just a larger view of the same system—rather than people on the landscape, consider towns and villages (collectives) that, due to their temporal proximity, will share a similar past and future—but not necessarily *identical*.

When a timeline is created, what has happened is that someone hired a temporal bulldozer and altered the *terrain*. With a TVG, you can target a specific feature of that terrain and with appropriate resonance, flatten a hill, change the course of a river, or build a mountain. You have not changed the course of anyone *living* in the temporal landscape, but now they have to work around the modified features of the landscape.

---

54 This is also how the EPR paradox works... *temporal adjacency*, not spatial adjacency, so regardless of spatial distance, the two photons are still affecting each other's orientation, like a couple of dancers doing a ballet in time.

Suppose you are driving down the road to work, but today someone dropped a tree across the road. Rather than deal with the alteration of the terrain, odds are you will just turn around, and take another route—an *alternate* route, or in the temporal landscape, an *alternate timeline*. Because the tree was in front of you, you consciously changed your future by taking a different route, and by taking that route, you will encounter many new things and situations that you would not have encountered on the old route.

Now here's the surprise consequence… there is *no temporal law* that says you cannot get out of your car, move the tree, and continue along the route you wanted to follow in the first place! It creates an inconvenience,[55] but is not insurmountable. Once you know what is coming, you don't have to still be in the way once it gets here. If you were standing on a street, looked up and saw a piano falling out a 10th story window right above you, would you just scream, "the end is near!" and get squashed, or just step out of the way?

If you remain ignorant, you'll just *follow the crowd* around the obstacles placed by those that formed the timeline, going where **they** wanted you to go. Let's face it, *people are lazy* and will usually take the easiest path. And that is how they keep control—providing easy "paths," not only in timelines, but in politics, legality, economics, food, fuel… just about anything you can name. Odds are you never even knew that there was a choice. Those who wake up, have a choice. Implementing the choice is another matter.

### The Illusion of Reality

In a communication with David Wilcock, he stated:

I had two different insiders tell me that the reality we experience is a consensus that is driven by our perception and thought, to varying degrees. They explained that it is a very classified secret that both *mirrors* and *magnifying glasses* can break through that. We can see a ghost, for example, in a mirror even though our perception will not allow it to be there if we see it right in front of us. Then the other insider said that when you put two mirrors facing each other it gets even more interesting—and that this was "very, very sensitive information."

You have to remember that most "insiders" *don't know*, as the experts they get their information from don't know either. Delenn's comment about news reporters is applicable here, "That which you *do know*, you *do not understand*. That which you *do not know*, you *invent*."[56]

I didn't "know" until I became an *outsider* and several of us former insiders started comparing notes and going "what the heck?" It took decades of research to get to the point where we have a viable model of what is going on.

There is a lot of confusion that arises with the concept of parallel universes or timelines. From what we've found, it would be better expressed as "potential universes" that are generated by a type of consensus reality. But you have to

---

55 Many of these timelines are "inconveniences", as Q'uo (channeled by Carla Rueckert) put it, regarding 2012.

56 *Babylon 5* episode, "The Deconstruction of Falling Stars."

understand that it is not a general consensus—it requires knowledge of how bioenergy works and some of the concepts promoted by Franz Mesmer[57] in the late 1700s. When people believe the "reality," it reinforces the pattern. This is why "social norms" are considered *so* important these days—a consensus reinforcement. If people thought for themselves, those potential realities would collapse.

Regarding mirrors, these techniques used to be referred to as *Faery Stones*, and for things like mirrors, it depends highly on the composition. The reflective surface must be a heavy metal, like the old silverbacked mirrors. The aluminum mirrors common these days (much cheaper) don't really work. The heavier isotope of silver contains a good quantity of captured neutrinos, which in a charged state can reflect a portion of coordinate time structure into coordinate space.[58] You can put two silver mirrors facing each other, one with full reflection and one partial, and you basically have a "ghost laser," where the pattern undergoes "light amplification" between the two mirrors as in the old laser setups. If people have a clear understanding of the "afterlife", those in power would lose their ability to control through fear—can't have that, can we!

### Project Camelot Video: Dan Burisch

I have some comments on the Project Camelot "Dan Burisch" video,[59] where he is talking about the *Orion Cube*, *Looking Glasses* and *timelines*.

The *Orion Cube* was the *core* of the Montauk Chair, used to direct temporal experiments. The cube is actually the navigation computer from an SM spaceship, probably one of the smaller scout vessels. Consider that in order for a spaceship to travel faster-than-light *without* time dilation, it must be able to successfully navigate through coordinate space *and* coordinate time, simultaneously. Essentially, the "cube" provides a window into both realms for the pilot to navigate with.

What happens with FTL travel is that as you pass the speed of light (the EM barrier), the spatial dimension in the direction of travel shifts into the coordinate time realm, as speed is a reciprocal relation between space and time—cross the

---

57 All the work on Mesmerism and etherology was debunked because it actually worked too well as a way to manipulate the masses, and the elite wanted to reserve it for themselves. An excellent summary can be found in James Stanley Grimes's 1850 book, *Etherology, and the Phreno-Philosophy of Mesmerism and Magic Eloquence, including a new philosophy of sleep and of consciousness, with a review of the pretensions of phreno-magnetism, electro-biology etc.*

58 An understanding of Larson's atomic model, plus the RS2 research done by Peret on quadrapolar neutrino charge, would be needed to comprehend how this actually works.

59 See: http://projectcamelot.org/dan_burisch.html

barrier, and you start having a temporal dimension instead of a spatial one. *But,* since time is 3D, not a 1D vector, you have to keep the ship going straight in the coordinate time realm, which means you need to be able to see the realm to properly navigate it. This is

*Project Camelot Video: Dan Burisch*

what the *Orion Cube* does. Upon transition to FTL velocity, the volume of the ship has been altered to 2 spatial dimensions and 1 temporal one—which means that in space, it appears as an *area*, not a volume, usually a flat disc or saucer shape, depending on the FTL speed. It gets flatter the faster it travels beyond the speed of light (the reciprocal relation: more time, less space). In time, it is cylindrical or jet-like, like a meteor streaking through the temporal landscape that needs an accurate flight path so it arrives at both the correct location in coordinate space, and in coordinate time.

In the *Phoenix III* project, they still had all the bits of the ship's navigational system, the cube, interface and pilot's seat—the Montauk chair setup described by Preston Nichols, Al Bielek and others. That's what we ended up hooking to the IBM mainframes, which were essentially a simulation of the responses of the spacecraft (though they didn't tell us that). They knew the ships traveled FTL, so they knew the device could somehow affect time and the project was built upon that principle. They were able to fool the navigation system into thinking the craft was accelerating past the speed of light, so it would open the temporal navigation window—the vortex—even though it remained fixed in space. Of course, that created a few other problems as occasionally happens with simulations versus reality, like the odd wall disappearing and strange behaviors of animals in the vicinity.

It appears they either had other cubes from other craft, or pulled the one from Montauk prior to the abandonment of the project in 1983. I'd bet they just replaced the room full of IBM mainframes with a PC these days—probably a lot more computing power than we had back then—which made the device much more portable. From what Burisch related, it doesn't look like things changed much—they're probably still using my original "chair" driver code I wrote for the IBM! I don't know what they did for the interface, as the chair required a person with psionic ability to use it. I'd guess they reverseengineered some sensors to pick up the general psycho-emotional activity in the immediate area, to control the projection into coordinate time (without some kind of "pilot,"[60] it would be random and useless).

The *Looking Glass* is *not* the same technology. It still has all the properties of the *Sun Disc of Mu*, so I still believe it was reverse engineered from that artifact. I assume they disbanded the Looking Glass technology *not* for the sake of humanity, but because it *stopped working…* the original Sun Disc, according to the Peruvian records of the Elder Race, made use of the naturally occurring gravitational null nodes (also called a vortex, "dead spot", or in conventional science, a "wormhole" or EinsteinRosen Bridge[61]). This allowed simple access to the coordinate time

---

60 Duncan Cameron was the best of the Montauk "pilots."

61 Another concept greatly misunderstood by conventional science, but easily

realm through the use of vibrational control (song, chant, mantra, instruments). But, as our conventional technology became more and more based on electromagnetism, with the corresponding strong EM fields of power distribution, it literally pushed these nodes off the surface of the planet—though they do continue to occur in the upper atmosphere and nearby space. Without a nearby node to lock onto, the Looking Glasses *stopped functioning*. Of course, if something were to interfere with the power distribution on a planetary scale, then these devices would once again start working.

Regarding the timelines, we are still on the Montauk-generated timeline, which is called T1V83. That has just about collapsed (the detour in the temporal landscape is rejoining the original road), along with the "Time Lateral" imposed by the Confederation,[62] many years ago. When both of those routes terminate, which should be by the end of this year, we will be back on the natural timeline—in other words, we'll be walking on the road to where evolution is supposed to take us, through 3-dimensional time—a *Timeline Zero* (T0). *Timeline 1* (T1) and *Timeline 2* (T2) are both artificial detours, with different groups erecting barricades and detour signs in the temporal landscape, trying to direct the population down those paths.

The *Timeline 1* detour signs are environmental disasters that are designed to increase the global temperature by several degrees. This is the New World Order preferred route. What the timeline is structured to do is to first increase the global temperature through the introduction of hydrocarbons into the atmosphere—not from flatulent cows, but something on a much larger scale, say something way out in left field, like puncturing the Earth's crust where it is very thin at the bottom of the Gulf of Mexico, to allow the almost limitless hydrocarbons (oil, gas) trapped in the asthenosphere to escape into the atmosphere and cause a massive greenhouse effect. Not that something like that could ever happen, right? The petrochemicals will raise the Gulf temperatures substantially, alter the natural course of weather systems in the northern hemisphere, and raise the global temperature 2-3 degrees C, which will actually give the northern hemisphere a richer growing season for a while—at the expense of the southern hemisphere, which will be ecologically devastated. Of course, land there will get really cheap and be bought up by the big Oil folks—how much of Chile and Argentina has BushCo bought? One more "accident" in the northern hemisphere to kick the global temperature up another 2 degrees, and the northern hemisphere becomes a wasteland with few residents, while the southern hemisphere will flourish—of course, all the powers that be will be nice and comfortable in their new, southern hemisphere paradise. Easy solution to get rid of the 3rd world "undesirables" first, then reduce the world population significantly (as the bulk of the population is in the northern hemisphere).

*Timeline 2* runs into a major storm in the temporal landscape, due to the solar

---

explained with cosmic sector (coordinate time) principles.

62 *The Confederation of Planets,* as mentioned in the *Law of One* material, also known as the *Divine Council* or *Andromedan Concil.*

transition.[63] Both T0 and T1 just catch the edge of the storm—T2 runs right through the middle of it. This is when the Earth experiences a "core flare"[64] that people have been interpreting as some kind of galactic or solar "microwormhole" event. It actually has to do with the solar magnetic ionization level, not wormholes, which will take a significant hit as our solar system crosses the galactic neutral magnetic sheet.

If you note in the T1 and T2 scenarios, time-traveling humans are present, designated P24, P45 and P52 by Burisch (the J-rod). Take a close look—both groups are still 3rd density and *no longer on Earth*. This should tell you something about what T1 and T2 are designed to do in the long term—to put mankind back into another 3rd density cycle, rather than to advance to 4th density—which is the route of Timeline Zero.

## The Realm of Coordinate Time

Back at Montauk, one of the early projects was to amplify a psychic's powers to control emotions and reactions of people at a distance. This was an accidental discovery, as the effect on the local people occurred in the opposite direction the SAGE radar dish was pointed in. The bulk of the radio energy was being reflected forward (towards the test subjects on a ship at sea), and what did not get reflected —and often ended up in town—had some peculiar properties to it. Both animals and people would pick up the emotions of the person in the chair and react to them as though there was someone there causing them. The only problem was that it was a nonlinear response.

*Everything* that went on in the chair room was recorded. The psychic couldn't burp without getting the attention of security. And they always had to tell what was on their mind and describe any strong

*The Realm of Coordinate Time*

emotions, as those variables would affect the experiments and we had no way of compensating for that. Through correlating that information with the behavior of people and animals nearby, they found that the emotional content was *temporally* displaced, shifted into the past or the future, depending if the psychic was thinking of the past (shoot, I forgot to return that library book) or in the future (wonder what I'll have for dinner tonight?) But the scale was larger. A thought a few hours ahead, like dinner, could translate to a week or more until the effect was manifest.[65]

Back then, they did not have a concept of coordinate time, but did have access to a "secret science" that they had apparently obtained from the Germans after World War II, which was based on some of the aether theories of the 19th century

---

63 See my paper on *Geoengineering*.

64 Peret, Bruce, "At the Earth's Core: The Geophysics of Planetary Evolution," *Reciprocity* XXVII, № 1 (Spring, 1998).

65 Larson explains this differential in his theory as the "inter-regional ratio," which is a normalization of coordinate time to clock time by scaling space.

(I recall Maxwell being mentioned on occasion, as well as Einstein's unified field—which he apparently *did* try to publish in the 1920s, but was pulled for one reason or another). But they did know that the speed of light wasn't a *limit*, as it is said to be these days, but a **boundary** that was the doorway to this aetheric realm, which somehow altered time.

It wasn't until after the project was shut down that Dewey Larson had published sufficiently on the nature of his *cosmic sector*—a realm of 3-dimensional time—that brought a lot of the pieces together. Looking back at all the anomalies, with that cosmic sector knowledge… things make a lot more sense.

When they were running the experiments with the chair back at the base, strange things would always happen. It got so frequent, that people just tended to ignore them. Things like sitting in the mess having lunch and having a whole wall just shimmer and fade out of existence for a few minutes, then solidify back to what it was. When stuff like that happened, people would just pick up their trays and move to the other side of the room and continue eating. Ghosts (people that you see but aren't there) and poltergeist-like activity were very common. Things would fly up into the air without warning, shoot across the room… made for an interesting job. They even had a team of paranormal experts from some psychic Institute studying what was going on and questioning folks. Every now and then we'd have to attend a presentation from them, probably to keep people from panicking and going public.

It wasn't all good stuff, though. I believe a lot of it was documented in the Philadelphia Project material. Now and then, rather than taking out a wall, it would catch a person or two, and they would "*go fast*," "*get stuck in the green*" or "*go blank*." Perhaps I should clarify some of those terms:

- When time begins to infringe upon space, first thing you sense is "going fast." It is a feeling that time is racing through you, though everything appears to be running at a normal pace about you. When you feel that, RUN. You're about to get "stuck in the green." I think it is nature's way of warning you to get out of the way.

- What happens next is the air turns *green*. Pea soup green. If you've ever been in a storm shelter when a tornado goes overhead, you'll know exactly what I'm talking about. The air turns green and starts to take on a "thickness" to it, eventually like trying to move through Jell-o. You can get out of the way when it first starts, if not, then you are "Stuck in the Green." *Time* is perceived in space as a *solid*—the inverse of the spatial vacuum—or an "aether" that has a thick soup feeling to it, like trying to run in a swimming pool.

- The green then fades out and the objects in the field start to shimmer like the heat waves in a desert do to things in the distance. The air doesn't shimmer—the objects do. It is as though waves are running through them. Only lasts a few seconds, then "goes blank." It becomes somewhat *invisible*, though it is still physically there, like a wall turning into a clear liquid. This is what they were originally trying to accomplish—optical invisibility on the DE 173 (USS Eldridge, *Rainbow Project*).

- When a person goes blank, you have to act fast and mark out their location and lay hands upon them, as they are not easy to see. Usually anyone new to the base would be commanded to do this, as it was found that exposure to these field effects was cumulative, and if you tried to save someone once too often, *you* would go blank as soon as you touched them. And you would have to touch skin-to-skin—if you could find their skin (when someone went blank in the winter, all bundled up with clothes, about the only exposed skin area was a person's nose—not easy to find). Though they were still there it was in a distorted form, so you'd have to feel around, and wait to see if anything happened. When you did make contact, skin-to-skin, they would start to transition back and you would normally call over others to help, and the person would come back to normal. Curiously, when inanimate objects would go blank, they would just come back after the field was turned off (as in the Eldridge). Only biological organisms needed an assist.

• If a person goes blank and no one is there to help, good chance they will "get stuck." Getting stuck means a person is *fully* conscious and *totally* paralyzed. You can't even blink your eyelids. It is as though clock time for the body has come to a halt, and not even the corpuscles in your blood can move through the bloodstream. People whom have been "stuck" go into a panic very quickly, for good reason. Those that have been stuck refer to it as, "HELL, Incorporated." It causes some substantial psychological trauma. Laying on of hands can get a person "unstuck" and back—though not always back to "normal." It is a terrifying experience.

• Depending on the strength of the field and the time you've been stuck, you can also "Freeze." The shimmering form disappears, and there is no physical evidence that anything, or anyone, is there. If the position was not carefully marked, there is no returning from the freeze. When a person goes into a freeze, the area is roped off and everyone is kept clear, save some specialists with demagnetized equipment they use to unfreeze someone. In this state, there is a high sensitivity to localized magnetism. If a person with a pocket compass gets too close, the person in the freeze will *spontaneously combust*—and will burn there for hours or even days. Those that have been brought back say that perception is immensely distorted and are acutely aware of the passage of time—minutes feel like days. You cannot comprehend your environment; it becomes ghostlike and distorted, like you are between life and death and both realms are mixed together about you. And you cannot move, talk or do anything about it.

• A person can also be pushed into a "deep freeze," where the sense of time is completely *gone*. It is impossible to tell if a minute or a century has passed. The few that have been brought back from a deep freeze say that there is another world out there, a world that defies description— and they are usually overcome by madness, a result of their inability to comprehend what happened. (We now suspect they have transitioned into the cosmic sector, where their physical and motor skills are useless, and everything would be seen by human consciousness as insideout.)

That's just some of the fun of messing around with coordinate time. We ran into a lot of problems trying to deal with the cosmic sector (Larson's name for coordinate time), as all interaction with it—no matter how precisely calculated—would have random consequences. As it turns out, coordinate time is not this empty void that the 19th century aether researchers led people to believe—it is an entire *universe* unto itself, with stars, planets and *life*.

The shamans of ancient traditions were fully aware of the aetheric life in this coordinate time realm.

They developed skills to actually see them and their interactions with people. Many of the lower life

*The Realm of Coordinate Time*

forms are parasitic in nature and are attracted by a person's *qi*[66]—tasty food! We eat food to build energy; they eat energy to build form. They are attracted by strong emotions, particularly the negative ones such as fear—which was abundant at Montauk. It was the presence of these temporal entities that messed with a lot of the calculations, though we were pretty much unaware of it at the time (if we realized it, we'd probably have called in an exorcist or two!) We knew something was going on that appeared random in nature (the movement of these aetheric life forms), but did not have a good understanding of the realm we were punching a signal into, nor that there might be non-corporeal forms living there.

In the Montauk literature I've read, it was said that Duncan Cameron summoned a creature from the Id, a monster from the unconscious. Well—our unconscious is the consciousness of coordinate time, since they are reciprocally related. What Cameron actually summoned was probably one of these aetheric life forms, akin to one of our great apes, and pulled it through sufficiently that it could directly interact with physicality and destroyed the base. Once all the transmitting equipment was destroyed (I think Nichols did that—I wasn't there that night), that entity moved back into phase with his own realm and disappeared from ours. But that kind of energy signature will leave footprints—people are still seeing some strange things out at "The Point"[67] from lobster boats.

## After 20 Years

As I've mentioned frequently, the people running the *Phoenix III* project really did not have a clue as to what they were doing. It was usually "trial and error," mostly error. They had some advanced technology that was billed as "foreign technology" to make you think it was Russian or Chinese—but even Russian and Chinese technology is based on the same physics that everyone is taught in school. This stuff did things that was "out of this world," and obviously it was. When they'd start talking

---

66 Qi is a Chinese term for bioenergy, also known as *prana* or *ch'i*.

67 "The Point" is a local term for Montauk Point, the end of Long Island, New York.

"foreign tech," those of us down at the bottom of the ladder would just look at each other with that, "yeah, anything you say" expression. None of us really had a clue of the larger picture, as we would only work on sections of projects. But with all the information that has come out since those times, a larger picture can be assembled from the pieces.

Some of the things that we did find is that there are two different kinds of "technology" that is in use. The electromagnetic technology we use today comes from the SMs. There is also a different kind of technology that is used by the enemies of the SMs, the LMs, or the "Little Men."[68] They have a mechanical technology that is similar to the "vibratory physics" people discuss with aether theory, in particular the research of John Worrell Keely.[69] The two technologies tend to be mutually exclusive; they stop working in the presence of each other.

There was also a great deal of difficulty with LM technology, as man does not have the physical senses to interact with it properly. You would pick up a rock and say, "weighs about 2 pounds." One of the LMs would pick up the same rock and say, "it's a B-flat." Their sensory organs work differently than ours. Our physical senses are more along the lines of the race we have a genetic similarity to: the SMs. We can utilize SM technology easily, but LM tech would be better relegated to singers and musicians, as it deals more with the cosmic / coordinate time aspect of things. SM tech is purely spatial, which is why it was the preferred technology for the Phoenix projects.

To understand these technologies, one must first be acquainted with the concepts of 3-dimensional, coordinate time creating a universe of its own, and that universe exists concurrently with our own 3dimensional, coordinate space realm. It is *not* a parallel reality or tucked away in some far corner of the universe—it is right here, right now, just shifted out of phase with our spatial reality so our physical senses do not detect it. However, our non-physical senses *can* detect it, and operate within it, which gives rise to "psychic" ability.

- *Precognition* is nothing more than seeing something in the distance, in the temporal landscape.

---

68 The LMs are another intelligence species native to Earth, abundant in legend and mythology as elves, sprites and pixies, referred to these days as *LaMerians* (not Lemurians—a French word that means the *people of the sea*), *water babies* on the west coast of America, or the *Nøkk* in Scandinavia (*Stargate SG-1*'s *Nox*, a race of peaceful ETs, are remarkably similar to the LMs).

69 Keely had a difficult time keeping his technology working, because he was building it during the industrial revolution— surrounded by incompatible technology. Also, the majority of current research into vibratory physics is operating under a misconception... as mentioned in the *Geoengineering* paper, they got it backwards! Most of the tuning required to get a vibratory device operational is to *neutralize* vibration—not create it. Like Keely, LM tech is based on the *neutral center*.

- *Telepathy* is two people standing next to each other in time chatting, regardless of how far apart they are in space.
- *Telekinesis* is just manipulating the temporal component of an object with your temporal arm, and watching how "time changes space."
- *Clairvoyance* is a pair of temporal binoculars.
- *Clairaudience* is yelling down the street at someone in the temporal landscape.
- *Clairalience* is a barbecue in the temporal neighborhood.

Virtually *all* of the extra-sensory abilities are easily understood,[70] once you realize that you exist in two different realms, a spatial, material one for the body, and a temporal, cosmic one for the soul.[71] And the funny thing is, they aren't "extra-sensory," magical or metaphysical at all—just a natural consequence of biological life, which we can either choose to learn and use, or ignore.

In closing, I'll pass on a little secret... think back to High School math class, and a rather annoying little concept known as an *imaginary number*. A number that acts like a rotation and does not exist anywhere in space as a quantity. From a young student's perspective, imaginary numbers are harder to deal with than fractions! From our perspective, time is a polar realm—rotation is what occurs naturally. The material and cosmic sectors are better described as *complex conjugates* of each other,[72] so *space is real* and *time is imaginary*—but not in the sense of "make believe," but in the sense of the imaginary number. Understand the *complex number*, a combination of real and imaginary, and you'll understand the connection between space and time; yang and yin; body and soul.

*Since you know that coordinate time isn't imaginary, those funny little imaginary numbers are actually showing an interaction between the physical and metaphysical... and with a foot in both realms, it does open the door to those things that are beyond space and time.*

---

70 Larson, Dewey B., *Beyond Space and Time*, various chapters.

71 People refer to the *soul* by different names; sometimes *mind, etheric/astral body* or *spirit*. The researchers in the scientific underground make a clear distinction, defining the *soul* as the *cosmic body*, the aggregate of cosmic atoms that are the unseen half of a life unit, of similar size and complexity as the spatial, physical body. The *spirit* is relegated to "beyond space and time."

72 Larson considered the sectors reciprocals, but Prof. KVK Nehru demonstrated in the 1980s that they were better described as *conjugates*, in order to preserve dimensiona

# EXTRA-DIMENSIONAL & EXTRA-TERRESTRIAL ENTITIES

## Introduction

This is not going to be a paper on "who's who" and "what's that." There is already plenty of information available on the various species of extra-dimensional and extraterrestrial entities. So rather than *who* or *what* they are, this paper focuses on *why* they are, a somewhat unexplored area that resulted from working with dimensional equations on Phoenix III. Namely:

> 1.      The *structure* of extra-dimensional life, discussing the concepts of *dimension* and *density* that define our biological structure here on Earth, projecting forward into higher dimensions.
>
> 2.      *How to interact* with extraterrestrials, given the vast differences between us, physically, psychologically, intellectually and emotionally, even if we are of the *same* density.
>
> 3.      The next stage of mankind, ascension to extra-dimensional status—what is going to happen when we take our next evolutionary step.

## The Yin-Yang of Time-Space

When studying the relationships of space and time in the context of motion, it becomes obvious that we are talking about the same concept that the ancient Chinese philosophers were referring to with their concept of *yin-yang*. From the realm of normal observation, *space* is point-based, linear (kick a ball, it rolls in a straight line) and *yang*. *Time*, being the conjugate of space, is plane-based, rotational and *yin* (clocks go in circles across its face).

When reading on metaphysics, the terms "density" and "dimension" are commonly used to refer to higher states of existence. For the most part, authors consider the terms interchangeable, for example 4th dimensional or 4th density both refer to the next stage "up" from mankind, currently sitting at the 3rd density/dimension. This is going to require some clarification, as dimensional structure is actually *fixed*, and it is what is active *in* those dimensions that creates the relative *density*, and as such, the *quantum ontology*.[73]

## Dimension

"A property of space; extension in a given direction."[2]

First word that jumps out is *space*. Now that we know about 3D time, we have to *expand* the definition of dimension to include the *temporal dimensions*. So let us generalize the relationship—space and time are related as *motion*, so we are actually talking about *dimensions of motion*.[74] And those dimensions have properties— *properties of motion*, not of just space or time.

In space, we observe three dimensions: length, breadth and height. Whereas space is only an *aspect* of motion, motion itself must have three dimensions for this to occur and the other aspect, time, must also have three dimensions.

Therefore, we can conclude by observation that there are *only three dimensions*.[75] Of course, minds are now racing, thinking this precludes 4D and 5D entities. Not the case. Is that *label* counting dimensions *of motion*, or the dimensional *aspects of* motion?

## Density

"The state or quality of being dense; compactness; closely set or crowded condition."[76]

Though the term "density" has been used in metaphysical context for centuries, it only became popular after the publication of it in *The Ra Material*.[77] In that context, it refers to a discrete grouping of *relative complexity*. *Third density* refers to the current level of complexity that our biological organisms exist in, which is the three dimensions of space.

To understand the concept of *density*, keep in mind the reciprocal relation between space and time: it is like a seesaw, more time, less space. As consciousness begins to *expand* into the temporal dimensions, it appears to *compress* and become more dense, in a spatial sense. But the physical dimensions remain fixed, due to the

---

73 *Ontology* is the study of the nature of existence, so a *quantum ontology* is a snazzy way to say "levels of existence." 2   Dictionary.com on "dimension."

74 Larson, Dewey B., "The Dimensions of Motion," *Reciprocity* XV, № 1 (Spring, 1986)

.

75 KVK Nehru, "Some Thoughts on Spin," *Reciprocity* XXVI, № 3 (Winter, 1997), section 9, "Dimensionality of Space" provides a mathematical analysis of why only three, independent dimensions exist.

76 Dictionary.com on "density."

77 Elkins, Don and Rueckert, Carla; *The Law of One* series.

coordinate space they exist in, so what happens is you get a significantly higher amount of "whatever"[78] in the *same amount* of space. *More stuff* in the *same space* equals the physical concept of *higher density*.

### Time And Relative Dimensions In Equivalent Space[79]

Coordinate time cannot be directly observed nor measured. We can only observe the effects on *how time changes space*. So how does "time change space?" There are two ways, one in the *macrocosm* as we would see from iron filings orienting themselves in a magnetic field, or in the *microcosm* at the atomic level. Larson refers to the latter as *equivalent space*, which is the spatial "equivalent" of the temporal motion of atoms and particles.[80]

As already mentioned, the yin of time has an intrinsic polar or rotational nature, so when we try to express time in the linear system of space, it is a bit like Cro-Magnon man first encountering the wheel, after spending his life dragging loads on sticks in a straight line. If you were going to Bedrock University and Prof. Albert Einstone, the local expert on temporal physics, gave a pop quiz with these two instructions, could you answer them?

> 1.      Draw a straight line that is 45 degrees long.
> 2.      Draw two lines at a 45 degree angle.

The first instruction is the problem with trying to express 1D temporal rotation as a 1D linear, spatial concept. The 2nd is the *equivalent space* solution. Does Einstone's quiz make you think enough to understand the problem?

#### The Yin-Yang of Time-Space

Mathematically, time can be expressed by an *imaginary number*.[81] Note that an imaginary quantity *works* just like a *real* quantity—it has a magnitude and is commutative in math. However, you cannot place an imaginary quantity directly on a real axis, nor a real number on the imaginary axis. They are mutually exclusive.

That is where equivalent space comes in, a 2-dimensional space that *can* express this yin quantity of time as a 2D spatial rotation. So when we go to measure the motion of time in equivalent space, all the dimensions are kicked up a notch.

In coordinate space, the geometry is determined by the number of spatial

---

78 "Whatever" works for particles, atoms, life units, the psyche, complexes... it is of general application. Multiple mechanisms are not required, once this reciprocal relationship is understood.

79 A word play on *Doctor Who*'s TARDIS.

80 In the Reciprocal System, atoms are a temporal rotations existing at a coordinate, spatial grid. Since we cannot directly observe time, we see the equivalent space as a hypersphere, projected into 3D as a sphere—a tiny, round atomic ball.

81 daniel, "Time and Timelines", section "After 20 Years."

dimensions:

1D: Line

2D: Area

3D: Volume

However, when we look at the effect time has on space, the equivalent space, the dimensions of *time* have this geometry:

1D: Area (expressed as 2 spatial dimensions)

2D: Volume (expressed as 3 spatial dimensions)

3D: Hypervolume[82] (expressed as 4 spatial dimensions)

Nature always expresses itself at the *maximum dimension*. In the macrocosm, everything we view has 3 spatial dimensions. 1D and 2D spatial structures make for good Science Fiction, but have never been observed anywhere in nature. One may claim that an electric field is 1D and a magnetic field is 2D, and that is *true*, but *neither are observable as a **spatial** dimension*.

The maximum dimension for the equivalent space of the microcosm is *four*. When counting spatial dimensions, there are three in the macrocosm and four in the microcosm—a total of *seven spatial dimensions* available to observation, which gives the 7-fold structure of our reality that is commonly recognized.

This dimensional structure gives rise to associated *densities*, based on the level of complexity contained in each dimension. The common structure looks like this, from a variety of sources:

| | Density | Larson | Latin | New Age | Common | Eastern | Evolutionary Focus |
|---|---|---|---|---|---|---|---|
| **M i c r o** | 7 | | | | Buddhaic | Adi | *Internal* Ethics, intellect, agapé, love, achieving the divine |
| | 6 | | *Undefined* | | Messianic | Anupadaka | |
| | 5 | | | | Mental | Atma | |
| | 4 | | | | Akashic | Buddhi | |
| **M A C R O** | 3 | Ethical | Animus | Spirit | Causal | Manas | *External* Physical evolution of body, mind and spirit |
| | 2 | Biologic | Anima | Mind | Astral | Kama | |
| | 1 | Inanimate | Corpus | Body | Physical | Sthula | |

One can see that there is a clear break between 3rd and 4th density, where the focus

---

82 The hypervolume of 3D time being expressed in 4D space has given rise to the field of *hyperdimensional physics*: physics as measured in Larson's *equivalent space*, rather than conventional space, to account for the extra dimensions.

switches from developing the physical container of the body in the 3-dimensional macrocosm (ego), to developing the internal attributes that we associate with higher states of consciousness and the reaching out to the divine (higher self).[83] The switch from *space* to *equivalent space* explains why this barrier is there, for when one crosses from the 3rd density and into equivalent space, they begin to have direct access to the *dimensions of time* and the mystical universe it symbolizes.

Larson defines his three *Levels of Existence*[84] as:

1.    *Inanimate* (particles and atoms; chemistry, defined by the time or space regions).

2.    *Biologic* (living organisms, defined by the *life unit*[85]).

3.    *Ethical* (ethical behavior, violating biological survival needs, defined by a *control unit.*)

These levels also correspond to the conventional mind/body/spirit structure, or as it is referred to in medieval Latin:

1.    *Corpus* (body), the inanimate, chemical structure, that is *either* spatial *or* temporal.

2.    *Anima* (life, mind or soul), the biological structure composed of *both* space *and* time.

3.    *Animus* (intellect or spirit), that ability of reason and self-sacrifice, that goes *beyond* space and time.

These first three *densities* **all** have *three dimensions of space,* and though they have an atomic presence in time, there is no consciousness in 3D time. Take note that in Latin, *animus* equates *spirit* with *intellect*. Keep that in mind as you read this paper.

### *The Interrelation of 3D Space and 3D Time*

In order to move beyond the three dimensions of space (the first three densities) and into the realm of equivalent space (and the temporal dimensions it represents), a basic understanding of the two, coexisting sectors of the universe is helpful.

Our everyday life exists in a *material sector*, comprised of 3D space and clock time. As a balance to this, there also exists a *cosmic sector,*[86] comprised of 3D time and

---

83 "Out" in the sense of reaching out into 3D time, the metaphysical region, by going inward within ourselves. Again, that reciprocal relation can be quiet revealing when understood in this context. *In* in space = *Out* in time.

84 Larson, Dewey B., *Beyond Space and Time*, chapter 5, "Levels of Existence."

85 The life unit, being a stable combination of material and cosmic atoms (matter and antimatter), tends to generate *helical* structures. The linear motion of space combines with the rotational motion of time, generating helices like DNA.

86 The term *cosmic sector* was used by Larson because it was identified as

clock space. Together, they form a very nice symmetry to existence, the symmetry of a universe of motion. But it is important to realize that they are not different "halves" of the Universe, but exist side-by-side like "parallel dimensions" that are 90 degrees out of phase with each other. There are a few ways to visualize this interrelation:

- The sine and cosine trigonometric functions are 90 degrees out of phase—when a sine wave is crossing zero, the cosine wave is at its extreme; when the cosine wave is crossing zero, the sine wave is at its extreme. The cosine would be the yang, spatial wave and the sine the yin, temporal wave.

- Geometrically, as the difference between points and lines of a 2D diagram, such as a triangle, where you can draw the triangle by connecting 3 points, or intersecting 3 lines.

- Geometrically in 3 dimensions, between vertices and faces. For example, a tetrahedron can be drawn by connecting 4 vertices, or intersecting 4 planes to make faces. This is the visualization that is the most helpful in the 3D macrocosm of the natural world.

*The Yin-Yang of Time-Space*

From our conventional reference frame, we see 3D space as "connect a dot." We identify locations, then connect locations to create pathways and geometric structure. 3D time, being *unobservable*, acts *between* those locations as a *force* or *force field*—time is the line between two points, or the face between three vertices that can expand or compress to move the relative positions of the spatial points, which we interpret as the pushing or pulling of electric and magnetic fields.

## The Cosmic Sector

Because of the reciprocal relation between space and time as motion, *everything* that we see in space has its temporal equivalent. If one were to move their consciousness out of the material, spatial sector and in to the cosmic, temporal sector, everything would appear inside-out.[87]

However, if you were born in the realm of 3D time, you would claim that folks living here in the material sector had everything inside-out, upside-down and backwards, because your consciousness would be adjusted to viewing *time* as *locations*, and *space* as *force fields*. It is all a matter of perspective.

## Extra-Dimensional Entities

Now we have the basics to understand extra-dimensional entities, the majority of which are *entities with a presence in 3D time*. This includes entities that are native born in the cosmic sector that have learned to access space, and entities in the material

---

the origin of *cosmic background radiation*.

87 Being a 3D system, all the vertices and faces would swap places. Cubes would become octahedrons, dodecahedrons become icosahedrons—you would hardly recognize anything.

sector to have obtained conscious access to the realm of 3D time.[88]

Consider a cosmic creature, a native-born temporal entity that has their physical structure in time, and therefore can only interact with 3D space as *force*—they are *invisible* to our normal, waking consciousness, yet since *time changes space*, we can still bump into things that aren't there and they can make things fly around the room without any observable cause. Ghosts, poltergeist and the like are all entities of this nature—entities with a structure in 3D time.[89] And yes, *we* appear as ghosts to the cosmic life in 3D time, as we cross the barrier in the other direction![90]

Two other situations arise for extra-dimensional life, where an entity begins to *consciously interact* with the *spirit complex*, going "beyond space and time." When material sector life develops sufficient *intelligence*, that intelligence provides the *modus operandi* for free will to follow the silver cord[91] across into the dimensions of *equivalent space*—the temporal dimensions. *Use your mind to open the door to time.*

When someone here in the material sector begins to access the temporal dimensions, we call it *ascension*. When a cosmic entity, living in 3D time, performs a similar growth of consciousness, then they are actually accessing their *equivalent time*[92] dimensions giving them access to the *3D spatial dimensions*. These cosmic entities that are crossing over from the 3D time to 3D space are the ones we refer to as *ascended masters*, *angelic beings* or advanced *spirits*. (Not to mention those cosmic aliens that appear as hyper-intelligent, pan-dimensional beings manifesting as white mice.[93])

| Density | Space Dims | Time Dims | Equiv Space Dims | Perceived Dimensions | Ontology |
|---|---|---|---|---|---|
| 1 | 3 | 0 | 0 | 1D | Rocks |

88 Known in the old days as sorcerers, mages and magicians. These days, *spiritual* people.

89 This is a simple explanation; there are actually 13 stages of transition between 3D space and 3D time for biological life, but that requires an understanding of the various "speed ranges" of accelerating past the speed of light across multiple dimensions. This is the origin of why 13 is a mystical number, particularly in the black arts that manipulate 3D time.

90 In states of meditation, one can often bring consciousness to that "cosmic ghost" and interact with the realm of 3D time. This forms the basis of *magick*.

91 Also known as sutratma or *life thread* of the antahkarana, that connects the physical body to the soul.

92 The same concept as *equivalent space*, but from the 3D time perspective.

93 Adams, Douglas, *The Hitchhikers Guide to the Galaxy*. And yes, I'm joking. Everybody knows that pan-dimensional white mice aren't hyper-intelligent.

| 2 | 3 | 0 | 0 | 2D | Plants & Animals |
|---|---|---|---|----|------------------|
| 3 | 3 | 0 | 0 | 3D | People |
| 5 | 3 | 1 | 2 | 5D | Ascended, Spiritual Man |
| 6 | 3 | 2 | 3 | 6D | |
| 7 | 3 | 3 | 4 | 7D | |

OK, who stole 4th density? Why was it left out of the table?

To make a point. Remember Prof. Albert Einstone and the challenge of expressing a rotational dimension in a linear system? There is no 1-dimensional form of equivalent space, so you cannot get a 4D structure directly from the dimensions of space and equivalent space. However, that *does not preclude* the existence of 4th density nor a 4D structure to life, courtesy of a little gem discovered by Leonhard Euler back in the 18th century, which is known these days as "Euler's formula."[94]

Without going into the mathematics of concepts like *dimensional reduction*, when we interpret spacetime as a complex quantity and substitute *space* for the *real* component and *time* for the *imaginary*, it shows that the first manifestation of motion (the ratio of space to time) shows up as *1-dimensional waves*. If you consider the region of equivalent space as a sphere with a bunch of "wheels within wheels" spinning around inside as temporal rotation, these 1D waves are ripples on the surface of the sphere. Sort of a "half dimension" of equivalent space, and since half of 2D is 1D, we have now found the missing 4th density:

| Density | Space Dims | Time Dims | Equiv Space Dims | Perceived Dimensions | Ontology |
|---------|-----------|-----------|------------------|----------------------|----------|
| 4 | 3 | ½ (*vibration*) | 1 | 4D | Vibrations of love and light |

This shows the dimensional structure of the material side of things, starting with 3D space. The flipside works the same way, just exchange the labels of space and time and you have the whole range of ascending, *cosmic* beings. These beings do not interact with our conventional, spatial realm until they begin to ascend to their *equivalent time* dimensions, bringing consciousness to their *interior*, which shows up in 3D space as our *exterior*—that reciprocal relationship again! These cosmic beings, through their meditations, can easily affect the arrangement and structure of matter here, in our material realm, and are therefore considered angels, demons or deities.

---

94 Euler's formula is: $e^{ix} = \cos x + i \sin x$.

## Extra-Terrestrial Life

For us regular folk, interaction with other life has been restricted to the creatures of Earth. With the exception of domesticated animals such as dogs, cats and horses, mankind seldom even interacts with the other life of Earth, except to swat the occasional fly or squish a menacing spider. We do tend to be an aggressive race and without doubt, are the best killers on the planet. Interaction with extraterrestrial life is now inevitable as we live in a rather crowded solar system. So when the Nox[95] come a-knockin' on your chamber door, what is going to happen?

Let's examine what might occur during our first interactions with an "alien" presence. And let's keep it simple and pick our first contact with one of our mythological friends that are very similar to *Stargate's* Nox, the Nøkk, a peaceful and philosophical race of water sprites that is rather familiar with *Homo sapiens*, as we used to interact quite a bit in days past.[96]

The Nøkk, or one of the races of the "LMs"[97] as they are known in the trade, are about the same scale of evolutionary development that mankind is, sitting on the $3^{rd}/4^{th}$ density transition. The big difference is that they are cetacean-like, not land mammals, so living in the depths of the ocean have created different methods of behavior and communication. Squeaks and clicks of aquatic life works fine to locate food or to yell, "watch out for that shark," but makes it rather difficult to discuss Descartes over sushi. As a consequence of their underwater environment, telepathy was developed early in the $3^{rd}$ density, though it is normally a $4^{th}$ density skill.

Consider the implications of a telepathic species. First, *no privacy*, so *no secrets*. No secrets, *no hidden agendas*—everything is out in the open. With no hidden agendas, a peaceful existence results from working towards common, evolutionary goals. Cooperation that is based on *rapport*, not the competition of *rivalry*.[98] This type of

---

95 The Nox are a peaceful, advanced race in the *Stargate SG-1* universe.

96 There are stories of "little people" in virtually every culture on the planet. Like most species, they have their good guys and their "Cabal," but the bad guys are insignificant by comparison to the human version. These two groups are identified by the Celtic *Seelie* and *Unseelie courts*, or *Ljósálfr* and *Dökkálfar* of Norse, Germanic and Teutonic legends, and by many other names.

97 The "LM" is an old gypsy term that refers to the general class of "Little Men" of mythology: the elves, dwarfs, sprites, faeries, Leprechauns, ... the list goes on and on. It derives from a colony of water sprites in ancient France, those "de La Mer" (of the sea), and later *LaMerians*, which is often confused with *Lemurians*.

98 A fascinating demonstration of the energy of *rapport* was done on a YouTube video by Lama Dondrup Dorje, as "A Discourse on the Heart Sutra,

telepathic network in the 4th density is known as a *social memory complex*, or SMC. Unlike the 2nd density counterpart of the *group mind*, the 4th density social memory complex *retains the individuality* of its members, including unique personalities and unique skills. It is basically the reciprocal of the group mind, having a many-to-one association of many entities consciously working towards one goal, versus the group mind of the one-to-many "queen bee," where one mind controls many slaves.

For a human, this can cause some interesting problems. Humans that *are* telepathic, and there are increasing numbers of them, usually engage in a "one on one" connection with each other. This is relatively safe and the psychological barriers are effective at blocking the probing of one mind to another.

However, when engaging a *telepathic species*, there exists a type of "telepathic ionization level" to the interaction. The concept is not that difficult to understand. Suppose you have the thermostat in the house set to 70° F and take an ice cube out of the fridge and put it on the counter. You also take the boiling kettle of water off the stove and set it aside. What happens? Before long, you have a 70° puddle of water on the counter, and cold tea, also sitting at 70°. This is a type of "thermal ionization."[99] Apply the same logic. If you walk into a room full of telepathic LMs, there exists a telepathic field that you will be exposed to, which will, in a very short time, pull *your* psychic skills to ***their*** *level*. One of two things will happen, depending if you're more the "ice cube" and have to come up to temperature, or the "kettle of water" (well developed psi ability) and have to cool down.

### *Muggles*[100] *Meet Albert Einstone*

In the "ice cube" situation, a person with little to no psionic skill suddenly hears voices in their head, thoughts that are not their own and a wild mix of emotions that have no correspondence to what you should be feeling at the time. This normally results in a state of panic and if the exposure is not removed very quickly, schizophrenia or insanity might set in from the sudden breaking down of barriers in the psyche. Interestingly enough, most of what a muggle experiences is *not* a "transmission" from the telepathic species encountered, *per se*, but the *contents of their own psyche* that can now jump the barriers the ego has spent years erecting. Those repressed contents suddenly realize they can be heard, want their say, and do it like a screaming child. The flood waters of consciousness just got too high, went over the dams, and there you are—face-to-face with your darkest truths, as

---

with Chi Kung." As he vividly shows, you can't compete with rapport!

99 Larson, Dewey B., *Basic Properties of Matter*, chapter 5, "Heat." Larson generalizes the use of *ionization*, having three forms: *thermal*, *electric* and *magnetic*. (Magnetic ionization is currently unknown by conventional science.)

100 Rowlings, J.K., *Harry Potter*, A non-magical person, used in this context as a 3rd density human, with no psi ability.

well as having to deal with a scary alien standing in front of you.

Many telepathic species are well aware of this situation and will normally *not engage* a non-telepathic entity in a group situation, simply as an act of compassion and understanding.[101] Historically, when the LMs were interacting with humans and a human started to have this reaction, they would immediately release them and depart, as not to cause psychosis. (For the most part, the Nøkk are a decent species, though for centuries they considered the aggressive *Homo sapiens* as more of a "rabid dog" than a fellow traveler. And just as we domesticate dogs, the LMs also "domesticated" some humans, usually seamen, whom they would snatch off boats, mid-ocean, leaving an empty ship floating around. But those sailors ended up having a pretty good life with the LMs, traveling the cosmos on the Arks[102] with a *greatly* extended lifespan.)

### Albert Einstone Meets the Nox

The other situation arising from "putting the kettle on," is when a telepathic species encounters a person with an inherently strong, or consciously developed psi ability. That person will actually *overwhelm* the social memory complex of the telepathic species. In this situation, the human psychological barriers tend to stay in place, but the flood waters of consciousness spill *out* of the mind and across the telepathic landscape of the social memory complex. Most telepathic species, like the LMs, adapt quickly to this and it gives them quite the "high," from all that extra bioenergy entering the complex. When this occurs, both parties experience a wonderful exchange of energy and information while retaining their identities. But, like all good things, when the bioenergy finally dissipates there is the inevitable "hangover." For the person interacting, their bioenergy (qi, ch'i or prana) will be severely depleted[103] and they will feel depressed and burned out, sometimes for weeks after the encounter.

### Physical and Emotional Responses

Those scenarios address the basic communication issues with a telepathic species.

---

101 There are those that will deliberately use telepathic ionization to break a person's spirit, therefore opening the door of the psyche to a "reprogramming." Rumor had it that this kind of research was carried out at Montauk, but I have no direct knowledge of it.

102 The LMs refer to their "motherships" as *Arks*, self-contained biospheres of substantial size, constructed from asteroids and small planetoids. "Dead Arks," ones that are no longer functional, can usually be identified by white surface scars, revealing an underlay of ice.

103 Research has shown that the experience depletes dopamine severely, and the levels stay very low for up to 6 weeks. The person then experiences all the symptoma of dopamine deficiency. L-DOPA supplements have shown marked improvement in recovery after the interaction with a telepathic species.

There is also a "physical challenge," as we are not accustomed to the sights, sounds and aromas of drastically different life forms. And if you check your legends on the faery folk, they are very human-looking to start with, with some subtle difference. The Nøkk, for example, have the physical stature of a human child, 8-10 years of age, but have the skin of someone much older. It creates a contradiction within the psyche that has a reaction much like *seasickness* (when your eyes say the boat is still, contradicting your inner ear saying your moving). Unconsciously, something does not "add up" and you get a little nauseous, because you are used to people with that stature having young skin, clean hair and a smile—and not having bluish-green, dolphin-like skin with pointy, barracuda-like teeth. If you saw that "thumbing" for a lift on the highway, you'll hit the gas and speed away, despite them being a peaceful, philosophical people and wonderful conversationalists. Humanity has a *lot* of built-in prejudices that we need to overcome, and most people aren't even aware of them until something brings them to the forefront.

However, this can be overcome by strength of will and *not* looking away! When encountering alien species, you need to do so *full sensory*. That way your mind can build an internal model of what they are *supposed* to look like, smell like and sound like. You need to add them to your internal "database." Once that happens, then you are able to *recognize* them in the future and the impact becomes less and less severe, until there is no impact at all.

And there will be an *emotional* reaction, as well. Usually starts out as fear, sometimes to the point of panic with the "fight or flight" mechanism kicking in. This is where *intellectual* development comes into play, as you can reason yourself out of fear by changing that fear into *intellectual curiosity*. Fear pushes apart, while curiosity brings things together in rapport and understanding. The difference between "WHAT'S THAT!!!" and "Oh, I wonder what that is?"

Something else to understand is that most people get their emotional cues from *body language*. When encountering a body that you've never seen before, you will fall back on *human* body language—what you know—which can be *totally wrong* when non-human species are involved. If you smile at a furry blue creature from Alpha Centauri, he may interpret that showing of teeth as a growl and assume you're going to eat him for lunch. First encounters can be *very* sensitive, so you cannot take anything for granted—and you had better understand what you *are* taking for granted![104]

These are the situations that arise when encountering a *friendly*, telepathic species. Encounters between non-telepaths only have to deal with the physical and emotional issues, which are usually resolved quickly. The remaining situation is what happens when you encounter a *hostile* telepathic species.

### *Mankind Meets Godzilla*

---

104 Desmond Morris' series on *The Naked Ape* can be quite informative in this respect, as it treats human behavior as nearly identical to the behavior of the great apes—quite revealing.

Hostile species, such as the SMs,[105] are *fully aware* of the situations described during our LM encounter scenario. Many are also telepathic and share a common goal— but that goal may well be *subjugation*, rather than the search for enlightenment. Good predator take advantage of situations, so they make use telepathic ionization, physical shock and emotional stress to press their advantage.

Not much can be done to guard against the fear brought on by slit eyes and a forked tongue; man has had an adversarial relationship with *reptilia* for some time, and for good reason.[106] It is a kind of "race memory," if you will, that warns us of danger when certain extraterrestrial species are present that *Homo sapiens* has encountered before. Not all saurians are bad guys, but in their involvement with Earth has tended towards controversy in the past.

Even if humanity removes the SM influence from Earth, we will encounter them as we move out into the galaxy, so it doesn't hurt *to be prepared*.

## Defense Against the Dark Arts
### *Rule #1: You cannot defend against the unknown.*

Know thy enemy. This section on extraterrestrials has introduced the concept of telepathic ionization as both a beneficial form of communication and a potentially hostile control situation. What has been explained is very rudimentary. Subtle forms of this mechanism are commonly known as *subliminal programming*, which is a prime marketing tool for products and services, as well as a tool of control by authoritarian figures. "Subliminal" means below your threshold of consciousness, so you react to it, without actually being aware of it. It is based on the *group mind* response, not the social memory complex. If you *do not realize* you are being influenced, you *cannot* defend against it.

SMs use these techniques to bias your "fight or flight" mechanism to *flee*, where you will run right where they want you to go. Particularly if you are with a group and outnumber them—they make you scatter so you can be picked off singly.

Upon sensing the impulse to run, an effective technique has been to use your conscious, free will to *override* the impulse and stand your ground. That situation causes a kind of "telepathic backfire," since they were focusing on you to flee, you reflected it back, and now *they* have the impulse to flee. If you outnumber them, a sudden turnabout of, "let's get 'em, boys!" and there is a good possibility that *they* will turn tail and run—and they have the tails to turn and run.[107]

---

105 The "Space Men" or "Saurian Men," as described by the gypsies. Commonly referred to as "reptilians," though they exhibit characteristics more of an amphibian, having fish-like tales, fin-like "wings" on their backs and a horny skull, much like the Oannes.

106 daniel, *Anthropology* paper (not yet released), describes the mythological interaction with the SMs on Earth and how they influenced the evolution of *Homo sapiens*.

107 The SMs are actually quite cowardly, which is why they operate behind

Note well that I am *not* suggesting you attack a 12-foot tall, sharp-toothed lizard. *But*, the unanticipated action will cause a moment of indecision that one can use to escape. And you really don't want to engage the SMs if possible. At least not unless you're at least a P-8 with some decent training from your friendly, neighborhood Psi Corps facility.[108]

### Rule #2: Never forget your Hoffman Lenses[109]

We are a human society and have grown accustomed to human behavior and body language. Take the time to learn about body language and the societal "norms" of your region. There are extraterrestrial and extra-dimensional entities that can mimic human form, using a variety of techniques. The most common of which are:

- *Psi-tri Projection*:[110] a three-dimensional, "psychic" image that is sent telepathically to those nearby that alters the impulses the brain receives through the optic nerve to give the appearance of something else, such as making a saurian appear human. Psi-tri is no longer used much, because cameras and video equipment cannot be influenced and will show what is really there.

- *Transmogrification*: a chameleon-like ability that allows an entity to physically change their body structure to blend in with surroundings. Since the change is *physical*, this is the most difficult technique to detect.

- *Holocam*: short for "holographic camouflage," a device used to place a false image around something, using holographic technology. This works well for sight, cameras and video, but is revealed by *touch*—the hologram has no atomic substance, so you cannot physically touch the image. Projections are usually very close to the shape and size of the actual structure, to minimize this realization upon accidental contact. Holocam will also show up using infrared cameras, because the *heat* pattern will not match the *visible* pattern.

---

the scenes using groups like the "cabal" do their dirty work.

108 An organization of telepaths in the *Babylon 5* universe, known for its ruthlessness and underhanded trickery. The comment is a reference to the Psi Corp commercial in the episode, "And Now For a Word."

109 Carpenter, John, *They Live* (1988 film). Hoffman lenses were special eyeglass or contact lenses that allowed one to see through the subliminal techniques used by aliens in the film. A film well worth watching.

110 A term used by Tom Baker in the *Doctor Who* episode, "The Face of Evil." The scientific underground originally had this lengthy Latin name for it, then a Doctor Who fan just happened to mention a "psi-tri projection" as a "false, 3D image" and the term stuck, because it is very accurate and a lot easier to pronounce.

- *Invisibility*: There are three techniques[111] that can make an entity invisible:

○ A variation of *psi-tri projection*, to remove the image altogether. This is much more difficult than altering a shape and usually only done by more advanced entities. But again, will show up on cameras.

○ *Shifting to temporal displacement*: since structures in 3D time are out-of-phase with our illusion of reality, they cannot be observed and can still interact with space via force fields. But they can be detected by a *magnometer*.

○ Shifting from 3D space to 3D *counterspace* (also known as *exospace*), a region of space that has negative coordinates. Human perception only detects positive spatial displacement. These are the things that go *woosh* in the night—they can be *heard*, but not *seen*.

### Rule #3: Intelligence Controls Instinct

When you "lose it," you "lose out." Do not be *willfully ignorant*. If you don't know something, make the attempt to learn it. Even if you can't figure it out, or don't remember it, it will sit there in the back of your mind—if that information is needed in an emergency, it *will* rise up to the occasion.

This applies not only to extraterrestrial and extra-dimensional encounters, but also to the process of human ascension. The more you know about how it works, the more your consciousness can utilize that information to accelerate your progress along the path.

We know how electric and magnetic fields can be produced by inorganic substances. Life, being organic, also produces an analogous field that is generalized by the term, *bioenergy* (also qi, ch'i, prana and a variety of other labels). People realize that life has intelligence, but never realize that *bioenergy also has intelligence*. You can train your energy just as you can train your body.[112] When you make use of *intelligent energy*, it can leap to your defense well before your physical body can even flinch.

Just remember, there are *two sides* to the reciprocal coin for life units: 3D space and 3D time. When you begin to use the features of 3D time, such as bioenergy, you are essentially *ascending yourself* to life as an extra-dimensional being.

### In Conclusion

Something Larson points out in *Beyond Space and Time*, is that "anywhere life *can* exist, life *does* exist." There is quite a variety of life out there and the life we see on Earth, on the surface, in the air, under the water and in the ground, is representative of about a quarter of what is "out there." In other words, "we ain't seen much yet." Our life forms here are limited to a very specific environmental range.

---

111 There may be more than three techniques; our researchers have only discovered these three.

112 Eastern practices such as meditation, *chi kung, nei kung, qi gong* and others develop this intelligence.

As we move out into the galaxy, we are going to encounter entities that are far from our environmental expectations. But what is important is that many of them are just other peaceful explorers, trying to make sense out of their lives, just like we are. Remember that we share more "microcosm dimensions" than macrocosm ones, so though we may look very different on the outside, we are a lot alike on the inside.

## Ascension: The Tomorrow People

Back in the 1970s, Roger Damon Price produced a Science Fiction series called, *The Tomorrow People*, that was about ordinary kids with extraordinary abilities— they possessed 4th density skills, such as telepathy, telekinesis and teleportation. This is where the human race is heading along the path we call *ascension*.

Since humanity hasn't "been there; done that" regarding ascension, there is *no hard data* on the ascension process, so it comes down to an *educated guess* as to what happens. But, with the knowledge of how extra-dimensional entities exist in a universe of motion, we can apply that very process to our own evolution and make a good, educated guess.

When we start to access the microcosm dimensions of equivalent space, we are also accessing the realm of *coordinate time* (3D time). People have been doing this for centuries, using contemplation, prayer and meditation. But since our environment was still "3D," it was a difficult process to attain and master, often taking many years of devout study. As our environment is upgraded, that process becomes significantly *easier* to attain—but still requires the *conscious effort* to do it. Just as you can repress bad memories, you can also repress psionic ability. It is not *forced* on you, as that would violate *free will*. Nor is it a handout. It is an *offer*, that you can refuse or accept.

As *many* people have noticed, the sun is brighter and hotter than it previously was. And some days, it is really intense—obviously, we are already *in* the beginning of the solar transition and the planets are already responding to it. Opportunity is here, *right now*—we've got that *offer* to upgrade. Accepting that offer takes man to the next evolutionary step beyond *Homo sapiens*, to that of the *Tomorrow People*. **Welcome to Prof. Albert Einstone's TP-101 Class**

Congratulations on your decision to become a *Tomorrow Person*. Welcome to the next stage of human evolution!

You may note that the *way* you perceive a few things is changing, particularly the changing perception of *clock time*. Now that you have access to the first *dimension of time*, you will notice that you are no longer a victim of causality and can begin to consciously select the circumstances you will choose to interact with, in your personal future. The muggles refer to this as *precognition*, but as you learn to use your temporal eye, you will see it is nothing more than looking through a telescope at the surrounding, temporal landscape. However, at this point you only have *monocular vision*—a single dimension of time—and therefore have not yet developed the stereoscopic vision needed to accurately judge "durations" in 3D time. This does take some "time" to get used to, as you become familiar with the temporal terrain and learn how far things are away, based on their relative sizes in the distance.

When consciousness is placed in the dimensions of time, it goes beyond simple vision. You will also begin to detect the temporal equivalent to the other physical senses: hearing, touching, smelling and tasting. Note that these are initially perceived as *intuition*, rather than *sensation*. And they will be interpreted by feelings, rather than thinking, until you learn to use your consciousness to bridge the right and left hemispheres of the brain so information can be shared quickly and accurately between thinking and feeling. After this is accomplished, you will discover that many psychological concepts have the same reciprocal relation as space and time, yin and yang, and the material and cosmic sectors. Thinking and feeling are two aspects of the rational valuing side of consciousness; sensation and intuition are the "sensors" we use to pick up coordinate information from 3D space and 3D time.

So when you start "remote smelling" the cafeteria across the street getting ready for lunch, please put your stomach growls aside until the lecture is over.

This new, 4th density experience will be confusing to the consciousness at first, but as long as you understand the temporal landscape behind the strange feelings and intuitions, it is not difficult to cope with. However, many new Tomorrow People will have difficulty when they fall asleep and the landscapes swap positions, relative to their point of consciousness.

Falling into REM sleep is analogous to your consciousness accelerating past the speed of light and moving into the realm of 3D time. When that happens, everything flips and the consciousness perceives your *waking* thoughts, feelings and memories as though they were a *dream*. Muggles, having no consciousness in the dimensions of time, treat the waking and dream states as *two*, separate things. For them, they *are* separate because there is no bridge connecting them. Now you have that bridge, your dreams will change significantly as your consciousness will stay linked back to the spatial mind. Firstly, you will remember much more about a dream than you have in the past, and it will take on a *living* character. Just as you can go outside and nail some boards together to change your spatial environment, you will now be able to do the same in the temporal landscape, and alter it—plant a temporal garden, and watch it grow. Both the waking and dream landscapes will begin to merge into a larger reality, where science and magick become two aspects of the same thing.

An important point to remember is that at the onset of this merging, you will have the tendency to treat the new information as *external* to your psyche—you will think it is coming from the outside. But consider the years you have spent as a muggle, with two, separate landscapes—most people can barely remember a dream, and if they do, it is a fleeting memory at best. As a result, you are *not familiar* with your *existing*, temporal landscape, so it will appear to be something foreign your psyche. Do not get pulled into this trap. Remember that initially, *everything in your dream is you*. So get to *know yourself*, and once you do, you will find windows to the realm outside the psyche.

The most important point I wish to impress upon you is that you will now begin to experience the energy of *rapport*. This comes from being in the same temporal neighborhood as your fellow students. As you continue to grow, you will discover that *rapport* will replace the competition of *rivalry*, and *curiosity* will replace *fear*, because you have the basis of *understanding all things*, so there is no need to argue or compete. Growth is the fastest when you share what you have discovered on your travels in the new, magical realm of 3-dimensional time. So when you make the *conscious* effort to engage the energy of rapport through positive cooperation, things naturally "come together" and all involved grow from the interaction. Choose *harmony* over *discord*.

You will, from time to time, encounter others that have a consciousness in the temporal dimensions that have made the free will choice to harm others. It is best to keep clear of them at this time, until you have become accustomed to your "special powers." Here in Bedrock, they are the saurians we call the Rockefellers. The muggles call them the Cabal. My recommendation is to first learn what you can do with your own temporal access, particularly to identify what *is you*, and what *is not*, so you cannot be tricked by temporal detour signs or the "get your free superpowers here" booths used by the Rockefellers. The path of ascension is best walked consciously and carefully, until you are sure of your footing.

I can tell by the way everyone is licking their lips, that the Bedrock Cafeteria must be open. I believe Wilma and Betty have volunteered to serve you today. I hope you enjoyed this introductory lecture, and again, congratulations on your new standing as Tomorrow People, and we'll see you next time.

### Ultimate Answers to Life, the Universe and Everything[42]

#### *Will we spontaneously evolve and find ourselves in a different body?*
Nature tends to *adapt* to changing circumstances, "baby steps" rather than big jumps. If you go along with the transition as it occurs, you'll probably have your same body, but it will begin to get healthier and the adverse affects of aging will diminish. You will also find new areas of the mind opening up, like being able to sense other life in nature, which will eventually lead to telepathic ability and other psionic skills. But always keep in mind this rule of Nature, "if you don't use it, you lose it." So if you choose not to make use of the "special powers" that you have access to, they will not develop and you can stay a happy muggle.

Should your body die during the transition, then upon reincarnation you'll find yourself in a different body. Personally, I like the "baby step" approach, because I really don't want to spend another 9 months in the womb waiting to get out, then have to spend years learning how to use another body again just to get back to where I already was.

---

42   Adams, Douglas, *The Hitchhikers Guide to the Galaxy*.
*Ultimate Answers to Life, the Universe and Everything42*

#### *Will we end up on a different earth?*
I don't see any reason why we should. The Earth may have been beat up a bit by the cabal, but it's only on the surface… they are more like an itchy skin rash, than a disease. Down deep, the Earth is still healthy. Remember that Earth is a *living entity* and would probably appreciate those that stayed around to help to get the old girl back to health.

#### *Will there be three days of darkness until the sun has completed its shift?*
Last time was three days. Since the increase in magnetic ionization level of the sun

is not linear, fewer elements become available for combustion during the transition each time around. So, "at most, three days" would be the reasonable answer.

Understand that the sun *does not stop* producing light and warmth. All the normal energy processes of the sun continue on, though the disc of the sun becomes dark. (Not totally dark, as there will be veins of red and yellow running across the face where thermal motion remains in the low speed range, giving a crackly appearance. The poles may also continue to produce light, because of the way the magnetic fields align.) The sun goes dark because the *thermal* motion of the photosphere has accelerated to FTL speeds, so light just moves outside the visible wavelengths.

Of course, if you don't know what is happening it would be a terrifying experience. Personally, I'm just going to kick back and take some video of the event. Nothing to worry about, since I know it is temporary.

I don't know *when* it will happen, as there is just not enough data available on the solar core and the amount of matter available in the solar system as fuel. In the Geoengineering paper, I have documented the precursors of the event, many of which are apparent now. There is a good chance that the burst of radio emissions at the onset will knock out all the cell phones, radio and non-cable television, so if that happens, get ready!

**Wouldn't the general public, currently kept unaware of ascension, panic or be traumatized?**

They certainly will be. And it's up to people like yourself that understand what is going on, to help out your neighbors.

Keep in mind that we have *free will*. And that can be used to promote our evolution, or retard it. Some people will simply not be ready for ascension and it would be wrong to violate that free will choice to force it upon them. Each must choose for themselves.

**What should people do before the timelines finally converge? Wait for the cavalry? Stay informed and inform others?**

I don't know a single Native American that is waiting on the cavalry! Don't see why we should, either. The original Montauk "detour" is already so close to the natural timeline that it can be considered "over and done with." That can be seen with the way things are changing worldwide, economically, politically, spiritually and scientifically. All sorts of new things are at hand, now that we don't have that pull from walking the difficult path around that temporal mountain from that detour.

We need to focus on where we are heading from here. The globalists have paved a couple of nice, easy walking highways in the temporal landscape, with their promises and free handouts… complete with GMO restaurants and police checkpoints at every exit. But we don't have to take those routes. Granted, it is a little harder to pave your own way, but if you want true freedom, it's the only sensible choice.

Virtually everyone has precognitive ability—the ability to see a distance in the temporal landscape. Open your eyes and pick where you want to go—create your own timeline. Grab your family and friends and head out to that unexplored, temporal territory and boldly go where no man has gone before. Then send someone back to tell the rest of us what's out there, to help us decide if it is our path as well.

**How do you think humanity will explore the universe post-2012? Using stargates? Will our updated bodies allow us to travel by using the time-space world?**

Stargates are fine for long-distance travel, but heck, we have not been skiing on *Olympus Mons*[113] yet. There's plenty to see right here in the neighborhood. And only minutes away at Warp 1.

One of the psionic skills that does develop is *jaunting* (teleportation). But it has limited range and usefulness, as you can only go where you have already been, since localization is done by imagery. That skill does improve as you reach higher densities and are able to astral project first, then yank your body along behind you, like baggage in tow. But we're not there yet.

Humanity would be "out there" *right now*, if he did not have the reputation of trashing everything in his path. Laziness causes more damage than weapons. Who the heck wants mankind around, throwing beer cans out the airlock with his polluting, noisy chemical rockets blasting through the neighborhood, leaving trails of nuclear waste behind him?

When we learn to be good neighbors, those extraterrestrial neighbors will invite us over to tea, and even send their chauffeur-driven saucer to pick us up. We need to learn to live in rapport with nature and our neighbors, then everyone and everything benefits. It is just that simple.

---

113 *Olympus Mons* is a volcano on Mars, the largest one in the solar system.

# GEOCHRONOLOGY
# HIDING HISTORY IN THE PAST

## Introduction

One of the advantages of being a subcontractor for "black ops" projects is that you often overhear the strangest things—things that sound like science fiction or a good Halloween story, but you soon learn are very serious topics and you need to keep your mouth shut, until you are well away from the situation. And when you are poking around in history with the Phoenix III equipment, a *lot* of unexpected things turn up. And so is the case with the origins of man. Jonathan Glassner and Brad Wright,[114] hold on to your hats… because you were a lot closer than you realized.

These papers discuss *anthropology*, the study of the origins and behavior of *homo sapiens*, developing a *radically* different world view that will not only make anthropologists scream in horror, but will make religious folks want to bring back burning at the stake. The proposed theory is a common denominator to a lot of other research, mythologies and doctrine. It is said that there is some truth in everything, but in this case, a *lot* more truth than anyone ever realized—just happens that a few things got "lost in the translation"[115] over the generations. And that is what this paper attempts to correct.

When Dewey Larson created his *Reciprocal System of physical theory*,[116] he clearly defined what he wanted to accomplish—to define the *physical* universe. He set out using basic deductive and inductive reasoning processes to achieve his goal. When finished, he had a very powerful "theory of everything" that could explain the smallest photon to the largest super-galaxy, except there was one problem… there were still things that existed and were observed in everyday life, that his theory *could not explain*. Things like biologic life, extra-sensory abilities and the realm of ethics. So he took all these concepts and threw them into his "think tank," removed everything that *could* be explained by his *Reciprocal System of **physical** theory*, then took an objective look at what was left. The result of those left-over bits became his book, *Beyond Space and Time*, which discusses the concepts remaining in that think tank that cannot be explained directly by his physics. *Beyond Space and Time* was Larson's last

---

114 Jonathan Glassner and Brad Wright are the creators of the popular science fiction series, *Stargate SG-1*.

115 See the books of Mauro Biglino on the literal translation of the Hebrew Bible.

116 *Reciprocal System of Theory* website: http://rstheory.org

book (he died before it was published) and after he removed the pieces covering biology, basic metaphysics and ethics, there was still stuff remaining in that think tank for future researchers to figure out.[117]

I am taking a similar approach with these anthropology papers, which is not about digging up old bones but an attempt to put together a theory of origins of mankind using religious, scientific and mythological data under the common framework of Dewey Larson's *Reciprocal System*, which has proved to be very effective in "explaining the inexplicable" over the last half century. It is my hope that pointing out some alternatives to *unquestioned* beliefs, we can take a similar approach with our mythological systems that Larson did with the physical universe—clean up the misunderstandings, take an honest look at what is left, and develop a theory from that premise as *natural consequence*.

And what is left in that "think tank" is going to be the *really* interesting stuff, for it will provide the opportunity to open an unexpected door to our future.

**Geologic History**

We've all been told about the Earth being billions of years old, with mankind not showing his CroMagnon face until about 50,000 BCE.[118] Unless you have got a TARDIS[119] or Bill & Ted's phone booth[120] parked in the garage, who is going to question that? I wasn't around back then and neither was anyone I know. So we just accept what we've been told by the "experts,"[121] as usual. Guess what... *everything you know is wrong.*

*Geologic dating*, also known as *radiometric dating*, gives us our geochronology that is based on radioactive decay rates. Sounds all well and good with one exception, pointed out by our old friend, Dewey Larson,[122] in his discovery that radioactive decay is actually a *temporal explosion*, an explosion in 3D time—*not space*—as conventionally believed. The rotational structure of the atom, existing in coordinate time, explodes and scatters its pieces around in 3D time. As our clock time proceeds,

---

117 *The Reevaluation of the Reciprocal System of Theory* website: http://rs2theory.org

118 *Before Common Era*, the year 1. Formerly known as "BC," Before Christ.

119 TARDIS is an acronym for *Time and Relative Dimensions in Space*, a time traveling spaceship from the BBC series, *Doctor Who*.

120 *Bill and Ted's Excellent Adventure*, MGM, 1989. Bill and Ted time travel through history in a phone booth, a play on the BBC series, *Doctor Who*, where the Doctor travels the universe in a police telephone box called the TARDIS.

121 *Expert*, definition: A person who knows more and more, about less and less, until they know everything about nothing.

122 Larson, Dewey B., *Basic Properties of Matter*, Chapter 24, " Isotopes ."

we just run into the bits and pieces of the atom that physics views as *radioactive emission*. *Same* location in *space* (the atom), but *different* locations in *time* (the emission). A rough analogy would be to take a bag of marbles (rotations in the atomic time region) and dump them out in a hallway. You dumped them in one instant—like an explosion—but as you walk down the hallway, you run into the marbles as individual pieces at different clock times. From a purely spatial point of view, it looks like you have the bag in your hand all the time, and a marble jumps out of the bag and onto the floor when you get to the position where it came to a stop in coordinate time.

When a large atom explodes in time, many pieces get scattered all over the coordinate time realm; some nearby and some quite far away. As a result of this distribution, the larger the explosion, the larger the error in clock time interpretation—what is known as the *half-life*. What science believes is *millions* of years, is in reality, only *thousands*. That consequence, alone, is enough to make most scientist's hair stand on end. But it is a *natural consequence*[123] of the structure of the atom proposed by Larson in his *Reciprocal System*.

The second bad assumption is that once an atom becomes radioactive, it continues to decay until stable. That is not necessarily the case. The atom only has to throw off enough rotation to bring it back into the *zone of isotopic stability*, which it does in a single, *temporal* explosion. The basic rotation of the atom is still intact,[124] so it can continue to aggregate particles, charged neutrinos,[125] that can build its mass back up to the point where another radioactive detonation is required to stabilize it.

According to the physics texts, Uranium-238 decays to Lead-206 in a mere 4.47 billion years. That is how the age of the Earth is calculated. Now consider Larson's explanation. The first time you see a

*Geologic History*

particle fly off U-238, the atom *has thrown off* **all the particles** it needed to, to become a stable atom again. Those particles are just scattered "down the hallway" across time, and actually have nothing more to do with the atom, itself. The atom goes back to behaving like a stable atom and eventually enough particles collide with it to bring it into the unstable zone. It explodes in time, again, and throws off *more* particles "down the hallway."

---

123 Unlike conventional science, the Reciprocal System postulates a theoretical universe based on motion, to which Larson *derives consequences* and compares to observation. Most conventional theories are the opposite—the theory is developed to specifically *explain* an observation, so many things are missed.

124 With the marble analogy, only a portion of the marbles are scattered on the floor—some remain in the bag. In the Reciprocal System it is called a *mass limit*. Your marbles can only weigh so much in the bag, before you have to dump some out.

125 The charged/uncharged state of subatomic particles is not recognized by conventional science, leading to more errors.

In the meantime, physicists are sitting around with their stopwatch measuring clock time, waiting for it to stop throwing off particles. Since the atom is exploding over and over again in 3D time—and they are waiting to stop running across particle debris down the hallway—they end up sitting around a *long* time, say 4.47 billion years, even though it only took a few thousand years to run across all the particles from the *original* explosion.

Those particles with a short half-life are the ones that don't make a very big temporal bang, so there is a good chance you will run across every "marble in the hall" and run out, before it reaches the zone of isotopic instability again. But the larger the atom, the less chance there is of that happening and the dating error becomes exponential.[126]

As we correct for these errors, we find that recent history is a bit more "recent" than we thought. With that new information, we can now make an accurate correlation to the records provided by mythology and various religious apocrypha; some of which provide enormous chronological detail.

**Calendars**

Our calendar is based on *rotation*, the rotation of the Earth around its axis (days), the moon around the Earth ("moonth" or month) and the Earth around the sun (years). Historically, different societies record their calendars in different ways. Some count days to calculate years, others observe celestial alignments to determine when a year starts and significant events (planting, harvest), but don't really care about individual days. These different systems are all translated to our modern convention of days, months and years.[127]

Our current, 365-day calendar represents the way rotations and orbits occur *now*, but was this always the case? In order to be that static, rotation or orbit could *not* have changed over the millennia. The mass of the sun, Earth and moon would have to remain constant, despite all the meteoric dust they accumulate every day, and the internal structures would also have to remain constant—indicating that nothing much is going on inside the cores… not a very logical conclusion, given the observations. It makes far more sense that the lengths of the day, month and year have probably *changed* throughout our history.

If the length of the year was different, millions of years in the past, who cares? *But,* if it happened only a few thousand years ago, when mankind was alive and well, populating the Earth, he just might have recorded those events in his legends and that could make a *significant* difference to our account of history, particularly in correlating dates from different cultures.

---

126 I have run some estimates based on a re-computation of beginning of the Cenozoic epoch (the dinosaur extinction), and that 65 million year value only came out to be about *75,000 years*, curiously matching the start of the 3rd density discussed in the *Law of One* material.

127 I did not include the *week* as a natural, rotational period of a celestial "something," because what it was measuring is no longer there. This is addressed on page 19, explaining "what does God need, with a starship?"

Delving into historical and mythological records one finds that this *is* the case… everything changes, the length of the day, month and year, and observations of these events are usually associated with global cataclysms.[128] These worldwide geologic events indicate that something has shifted and a new Epoch, complete with different lengths of days, months, years, and climatic change, has begun. But there is something you should first understand about modern accounts of ancient civilizations:

My grandfather once told me a story about archaeologists, out digging in the remains of an ancient city that was the home to a pagan people. Some of the people still lived in a nearby town and they had some local helpers to clear away the dust and debris of the centuries. At the entrance to many of these homes they found a "blessing bowl," a small bowl that the residents used to sanctify themselves as their entered their homes, very similar to the Catholic practice of dipping ones fingers in a bowl of Holy Water and making the Sign of the Cross before entering a church. This was a major discovery, since it told much of the religion of these ancient people.

After a few weeks they began to run low on food, so they accompanied one of the locals to a nearby town to resupply. While walking through the street to the marketplace, they noticed those same, little blessing bowls by the entrances to the homes. Out of curiosity, one of the archaeologists asked their guide if they still practiced that pagan religion of centuries ago? The guide looked puzzled at first, looked at the bowl and replied, "No, but the dog still gets thirsty."[129]

With that in mind, let's ignore what the experts on the Mayan civilization have told us about the calendar and consider the words of an elderly K'iche' Mayan I met on a bus on the way to Chichen Itza, when he offered to explain the Calendar Stone I had on my T-shirt. I have long since forgotten his name, so we will just call him, "Bob."[130]

---

128 Cataclysms such as earthquakes, floods, torrential winds and meteoric bombardment.

129 Story from Bruce Peret, by his grandfather, Joseph Petrone.

130 With all due respect to *The Church of the Subgenius*.

Bob told me that at the start of the *human world*[131] there was but a single calendar and count; what we know these days as the *Tzolk'in*[19] and *Long Count*. There were 20 days in a month and 13 months, making the 13th the *last* month. "13" was considered an "end number," used to indicate the end of cycles.[132][133][134] When man was created, there were only 260 days in a year and the moon orbited the Earth in 20 days, not 28.

When I asked about the *Haab'*,[135] Bob said that it did not come into existence until the end of the 4th Baktun, after another great cataclysm[136] that moved the land and water about, fire rained down

*Aztec Calendar Stone*       from the sky and no one knew where anything was any more. The sky had so darkened that the sun was not visible for 20 years. After that,

*Calendars*

the gods gave them a new calendar, the Haab', having 18 months of 20 days to match the heavens. Out of respect for Ahau, they kept the Tzolk'in along with the Haab' and the dual-calendar system was born. Important to note is that there were only 360 days in a year at this time, for the 5-day month of Uayeb was not added until Huracan became angry and added the Uayeb (translated by Bob as "5 evil days") at

---

131 The zero date of the Long Count, 13.0.0.0.0, is actually the end of the 13th Baktun of the prior Age. The next day,

    132 .0.0.0.1, was the first day of the new Age.

133 The sacred calendar meaning "the division of days," that was provided by the god *Ahau* (also known as *Ah K'in*, from where the day name, *k'in*, originates). *Ahau* is also *Anu*, the Sumerian god, *An*, from the other side of the planet.

134 Much like the common, journalistic and editorial practice of using "30" to indicate a story is finished, originating from the last day of the month when it was due. Long counts that extend beyond the Baktun are often filled in with 13's, to indicate the end of the prior epoch: 13.13.13.13.13.13.0.0.0.0 was December 21, 2012. (Some say Dec. 23).

135 The civil calendar, originally 360 days.

136 This was the 2nd cataclysm that occurred at the end of a Baktun. The first was a great flood at the end of the 3rd Baktun that parallels the Hebrew Deluge. The start of a new Baktun was getting a really bad reputation for disasters.

the end of the 6th Baktun.[137]

Fortunately, the Maya used the Long Count to count days, so we know exactly how many *days* have passed since the creation of the human world. And we can adjust the number of years using Bob's calendar information. Shorter years early on mean *more of them*, so the calculated start of the Mayan Long Count was **not** 3113 BCE, but a bit further back... some 5773 years ago: 3761 BCE. Anyone familiar with the Christian Bible or the Hebrew calendar may recognize that year: *the year Adam and Eve were created*. A perfect match to the Mayan start of the *human world*.

Now that we have a very close correlation between the Mayan and Hebrew calendars, from opposite sides of the planet, other information becomes available. Hebrew accounts say the Great Flood occurred 1656 years after Adam. Converting that to a Long Count with the corrected calendar puts us near 2.19.16.0.0, which is about 6 years short of the end of the 3rd Baktun, the date of the Mayan flood. The adjusted date for the Hebrew Exodus from Egypt is about 1550 BC, with its plagues, volcanoes, earthquakes and the darkening of the sun. The end of the 4th Baktun was 1548 BCE.

Almost a 6,000 year period of history and the stories of the Deluge and Exodus match up with their Mayan equivalents to within 6 years? I'm sure that must be "coincidence"...

We've decoded the blessing bowl at the doorstep, so on to the "plates" in the kitchen.

**Growing a Planet with Expansion Tectonics**

The current geometry of the Earth is an oblate spheroid (flattened ball) of fixed dimensions, on which exist continental plates that slide around and bang into each other, an inch or two a century, creating mountains, valleys, earthquakes and volcanoes. The "science" is called *tectonics*,[138] and is the result of the standard, scientific approach of trying to create a theory to explain observation.

Let's take Larson's approach and use the concepts of the Reciprocal System to *determine* the inner structure of a planet, and find the *natural consequences* of that structure.

An initial draft of this research was published in Peret's 1998 paper, *At the Earth's*

---

137 Extrapolating, March 8, 748 BCE, Julian, or February 28, 748 BC, Gregorian calendar.

138 Tectonics has now been extended to: extensional tectonics, thrust tectonics, strike-slip tectonics, plate tectonics, salt tectonics, neotectonics, tectonophysics, seismotectonics and planetary tectonics.

*Core: The Geophysics of Planetary Evolution.*[139] It proposes a planetary model that is based on the remnants of a white dwarf star, the "B" component of the common red giant/white dwarf stellar binary. The idea is explained in detail by Larson in *Universe of Motion*,[140] where the "A" component, the giant star, reaches its age limit early and explodes in a supernova. What Larson failed to consider, and what Peret points out, is that the "B" white dwarf, sitting in close vicinity to a supernova, is *unlikely* to survive the explosion and will be destroyed in space and accelerated into time, producing a number of small, superdense fragments that evolve into the cores of planets.

Now that we know we're sitting on a white dwarf *core*,[141] Larson's research on white dwarf *stars* provides a road map[142] to understanding what is going on beneath our feet. The Reciprocal System demonstrates that the same processes occur over and over in nature, just at different scales, so knowledge obtained from quasars (galactic implosions) can be used with white dwarfs (stellar implosions), planets (dwarf fragments), asteroids (smaller dwarf fragments)... all the way down to atoms and particles, which are also just *motion in time, located in space.*

The planetary interior, like its stellar parent, is divided into four, distinct layers, based on the concept of motion, and how many dimensions of that motion exist in either space or time—what Larson refers to as the astronomical "speed ranges."[143]

---

139 Peret, Bruce, "At the Earth's Core, The Geophysics of Planetary Evolution," *Reciprocity* XXVII, № 1 (Spring, 1998).

140 Larson, Dewey B., Universe of Motion, Chapter 7, "Binary and Multiple Stars," pp. 83-102.

141 The analogous astronomical object would technically be a *pulsar*, which is a white dwarf that has been accelerated in time into the ultra-high speed range. The term "white dwarf" is used to refer to the general class of the star, but the *inner* core exhibits *pulsar* properties.

142 See: *Universe of Motion* , Chapter 6, "The Dwarf Star Cycle," and Larson's paper, "The Density Gradient of White Dwarf Stars."

143 Larson uses unity (1.0) for the speed of light and defines his speed ranges based on the idea that the *default* condition for everything is to *move at the speed of light. Gravity* is the *opposition* to that movement, so all his measurements are from the speed of light, downward, which he designates as "1-x" (start at the speed of light, and slow down to *x*).

SiAl,　　SiMa　　Crust
Mantle (1-x)
*Inward in Space (Gravity)*
Inner,　Inner　Core　(1-x)
Inner Core (3-x)
*Outward in Space (Antigravity)*
Outer Core (2-x)
*Inverse Density Gradient*
Each of these "speed zones" has specific attributes
that contribute to the behavior of the planet:

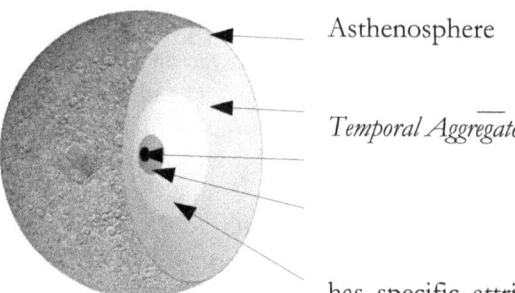

Asthenosphere

$\overline{Temporal\ Aggregate}$

1.　　**1-x**: *Low-speed* range of normal matter that comprises the Sialic (continental) crust, Simatic (ocean bed) crust, asthenosphere (slippery, magma layer) and mantle. The mantle is part of the original dwarf star remnant, whereas the crust is composed of meteoric aggregate. The mantle exhibits the property of *inward motion in space*, that we call *gravity*.

2.　　**2-x**: *Intermediate speed* range, where *two* dimensions of motion exist in space and *one* in time. This forms the outer core and has an *inverse density gradient*, where the shell of the outer core contains the densest materials, with a light, gaseous lower region. Since gravity requires all three dimensions of motion to be inward in space, and the outer core only has two, it exhibits *neutral* gravity.[144] Because of this presence of motion in time, the outer core produces an intense, scalar magnetic field.

3.　　**3-x**: *Ultra-high speed* range, where *one* dimension remains in space, with *two* in time. This forms the inner core and exhibits *anti-gravity motion*, as well as other properties associated with the *pulsar*. It is the balance between the gravitation of the mantle and the anti-gravitation of the inner core that keeps a planet in a stable orbit, much like trying to go "up" the "down" escalator,

*Growing a Planet with Expansion Tectonics*

at the same speed, and end up going nowhere.

4.　　**1-x**: *Inverse low speed* range, where *all* motion is in 3D time. This is the *inner*, inner core that was only recently discovered by geophysicists and named by Mehran Keshe as the *Caroline core*.[145] This appears as an empty, bubble-like void at the very center of the inner core that creates a link between the spatial and temporal structures of the planet, and when considered in a living aspect, would form the *soul*[146] of the world.

---

144 Larson, Dewey B., *Beyond Newton: An Explanation of Gravitation*, North Pacific Publishers. Gravity requires three dimensions in space. Motion in the intermediate and ultra-high speed ranges only have 2 or 1 dimension in space, and therefore exhibit neutral, or anti-gravity motion.

145 Keshe, Mehran, *"Static and Dynamic Plasma Reactors,"* Keshe Technologies.

146 The *life unit* in the Reciprocal System is an aggregate of material atoms

Because of the faster-than-light motion involved in a white dwarf star, the core exhibits conditions that are the *opposite* to a normal star.[147] Rather than increasing in density and heating up, the white dwarf decreases in density and cools down, creating an inverse density gradient[148] in the core of the planet. Due to the drop in density, the core *expands* over time, cracking the mantle and the crust sitting on top of it, creating the observed tectonic plates.

However, like their stellar counterparts, the white dwarf cores do not just slowly expand, they expand in discrete stages analogous to the solar transition, remaining quiet for centuries then when reaching a critical level of compression, fracturing the mantle, expanding the planet and splitting the crust along the tectonic fault lines. Because of the increased surface area, those plates are free to slide around on the asthenosphere,[149] a slippery magma layer that exists between the crust and mantle. The plates will then come to rest at their least energy configuration, which is usually a 90° rotation, bringing the massive weight of the ice caps to the equatorial region, like adjusting the clothes in an outof-balance washing machine. Note that the mantle, inner and outer cores have *not moved*; the crust just slides relative to the magnetic poles, so it *appears* that the poles are bouncing around on the surface.[150]

A simple way to understand this expansion is to take a balloon, inflate it, then cover it with mud and let it dry. Now inflate the balloon some more and watch what happens. The cracks in the mud become the ocean bed; the separate pieces of mud

---

(spatial, corporal) and cosmic atoms (temporal, soul or mind). Anything that is an aggregate of 3D space and 3D time, like a *planet*, is considered to be a *organism*.

147 Since astronomy has stellar evolution backwards, the white dwarf behaves like a "normal" star for astronomers, which is why they changed the main sequence stars to be "dwarfs" to account for observation. Unfortunately it is a common practice in science to change the observation to fit the theory.

148 Larson, Dewey B., "The Density Gradient in White Dwarf Stars."

149 The asthenosphere is also the source of crude oil, which is a waste product of a bacteria that lives in magma discovered during the Mount St. Helen's eruption. Most oil fields are crude that seeps up from the asthenosphere, so oil fields will eventually refill over time. An unlimited supply of oil exists just a few miles below the surface. The most accessible point are where the crust is thinnest—offshore—you know, where those thousands upon thousands of "deepwater horizon" drilling platforms are. And yes, you are running your car on bacteria poop, not decomposed dinosaurs.

150 Noone, Richard W., *5/5/2000 Ice: the Ultimate Disaster*. Noone had the right consequence, but the wrong cause. If the polar ice were to *melt*, by natural or artificial means, the crustal shift would be *minimal* during an expansion event.

become the continents. And as the balloon gets larger, there is more room for those chunks of mud to slide around on, giving the appearance of plate tectonics.

So as a natural consequence of planetary design, we find not a static sphere, but planets that increase in size with time, in sudden steps that rearrange their surfaces, accompanied with a lot of volcanic activity and usually resulting in a crustal rotation. These features are observed on the Earth, planets and moons.

In our ancient history, the planet they describe was *physically smaller*, the oceans were not as broad as they are now and the continents were arranged differently. In the earliest of days, prior to any oceans, all the continents fit together like puzzle pieces in one, small, hardened-mud ball—Pangea was an entire planet of dry land prior to the expansion, not a super-continent on a waterworld.

### Ancient Cartography

With what you know now about the size of the planet changing over time, look at some of the maps of the ancient

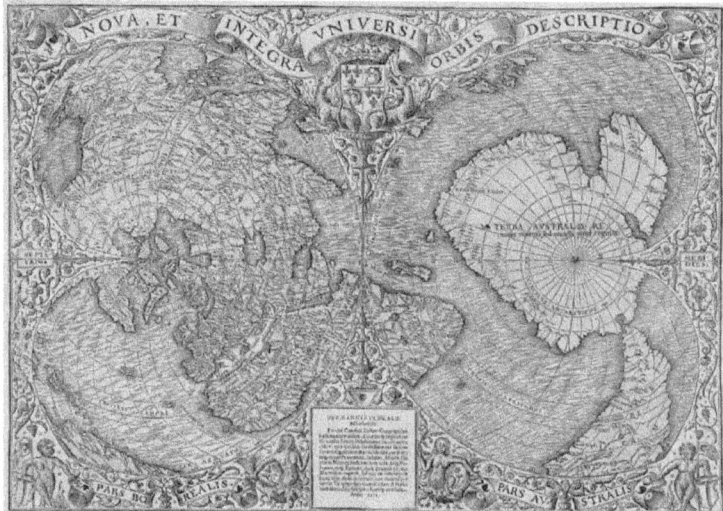

cartographers:

### Orontius Finaeus Delphinus (1531)

Note that Antarctica extends up to the Tropic of Capricorn and the map shows mountains on the continent, indicating minimal ice caps. The oceans are substantially smaller, and Italy is almost touching Africa.

The continental arrangement makes no sense now, but reduce the size of the planet and it tells a story of ancient times, of a smaller world with easily traversable seas, land bridges between continents for the migrations of peoples, and an entirely different climate than we have now. And this is not the only map of those times to indicate such a structure. Maps like these are found all over the world, indicating *common knowledge* among seafarers. Notes from the ancient mapmakers say that these were copied from even older maps, and how Christopher Columbus knew there was a "New World" out there to find again—he just did not realize the Earth expanded, and the oceans were a lot *wider* in 1492 than when the maps were originally made.

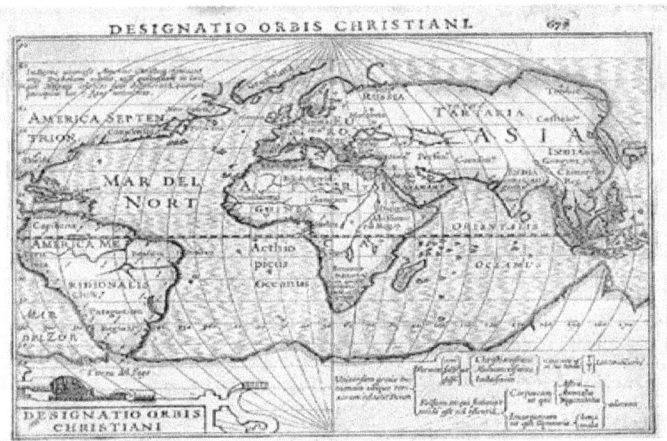

*Jodocus Hondius' world map first issued in the Mercator-Hondius Atlas Minor in 1607. This is one of the earliest thematic maps, featuring symbols illustrating the Christian, Moslem and idolatrous regions of the world.*

*Again, showing South America **attached** to Antarctica, as well as Australia being a **peninsula** of Antarctica. Why would so many ancient maps show this continental arrangement?*

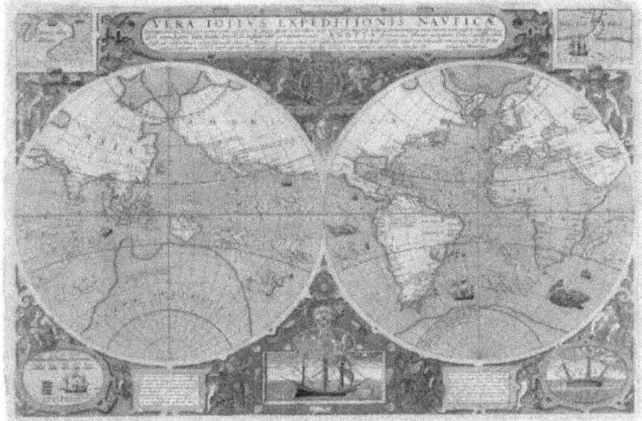

*Abraham Ortelius' elaborate double-hemisphere world map records the first English circumnavigation of the globe by Sir Francis Drake (1577-1580), as well as that of his countryman Thomas Cavendish a few years later (1586-1588). The map portrays the outlines of continents leaving the interiors blank, suggesting that the land areas were left unexplored. The marginalia includes the Elizabethan coat-of-arms, a vignette of Drake's ship the Golden Hind, and four corner illustrations. The drawing in the upper-left corner shows Drake's landing at Nova Albion in present-day California.*

**The Early Structure of the Solar System**

Fortunately, the gods provided us with some detailed descriptions of the early days: our *mythology*. Knowing what we do about geochronology and the structure of the Earth, these mythological records take on a different meaning—one that tends to fit the natural consequences of our theoretical development. In order to understand what mythology is describing, a more detailed picture of the early solar system is needed—one that is based on natural, evolutionary consequences.

As mentioned in my *Geoengineering* paper, one of the most important discoveries Larson made from the natural consequences of his Reciprocal System is that astronomy is *backwards*. Astronomers work with snapshots of the Universe, and they lined up their Polaroids from tail-to-head, rather than head-totail, and then tried to make sense of it. So the natural flow of evolution of planets, stars and galaxies was completely missed.

Larson's stellar evolution sequence proceeds from dust, to a red giant, orange giant, then on to the main stream and up to the blue supergiant, then a Type II supernova explosion (thermal limit),[151] following the spectral class sequence of N, R, M, K, G, F, A, B, O. But unlike conventional astronomy, stellar evolution *does not stop* with the supernova, because the supernova explodes its outer shell into *space*, and the inner core into 3D *time*. Whereas an explosion in time is analogous to an implosion in space, what you end up with is a super-dense, invisible object, emitting X-rays, surrounded by a large quantity of dust and debris concentrated in a ring (intermediate speeds): the black hole and its accretion disk.

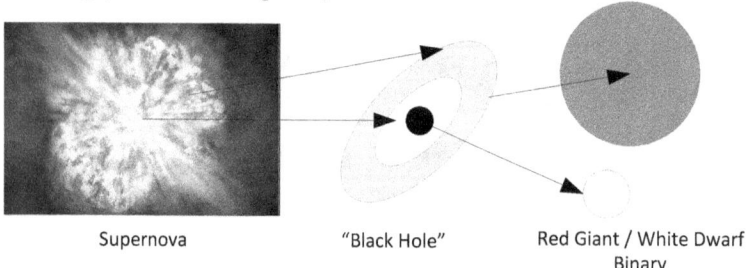

| Supernova | "Black Hole" | Red Giant / White Dwarf Binary |

Now we know the mechanism, we also know that this "black hole" isn't a hole at all. It is just the *spatial location* of a *temporal explosion*. As spatial gravity pulls the dust and rock together to form a new, red giant star, temporal gravity will pull the black hole together in time, expanding and cooling it in space, moving it from X-rays to visible light, resulting in the formation of the very common "red giant / white dwarf" stellar binary. And the situation can repeat, *ad infinitum*, producing stellar triplets, quadruplets, quintuplets, etc., or stars with solar systems.

Larson refers to this recursion of stars as "generations." The first generation is a

---

151 The "Type II" supernovae occurs as a consequence of the thermal limit—the star gets too hot to hold itself together. Type II supernovae originate from blue, Class O supergiants. "Type I" supernovae are the result of the atoms of a star reaching the "age limit" of matter, which can occur in any type of star. The controlling factor here is how old the fuel (matter) was that the star pulled in to consume in fission.

star that has not yet become a supernova, being found primarily in globular clusters. The second generation is the stellar binary. The third generation can either be a stellar triplet or a single star with a solar system.

The formation of a planetary system during the 3rd stellar generation occurs when the non-white dwarf star of a binary, what Larson calls the "A component," explodes in a Type I supernova. The energy accelerates the white dwarf companion into ultra-high (pulsar) speed ranges—anti-gravity, so it begins to move away from the A component, then was broken up into a number of fragments. Should the white dwarf component explode instead, the mostly gaseous "A component" would just be splattered around, to reform into a star, producing a stellar triplet.

*As we saw earlier, the two linear units from zero to the one-dimensional limit correspond to eight three-dimensional units. The constituents of the white dwarf are thus distributed to a number of distinct speed levels, with a maximum of seven.*[152]

When Larson refers to "speeds," he is talking about a *scalar* speed, not a *translational* one. Scalar motion increases or decreases in *integer steps*, not a smooth transition. So any internal, scalar speed of a fragment, $2 <= n < 3$ would be a speed of 2. There are no fractional parts. Because of this quantized separation, the fragments fall into tight, discrete orbits and the pieces tend to consolidate in those orbits, into one, large chunk we call a planet. This is unnoticed because of all the initial debris from a supernova explosion, and by the time the "dust settles" you just have a single planet in each discrete orbit.

But also notice that Larson states a speed limit of "eight, three-dimensional units" that are linearly distributed from zero to seven. That means that there are just *eight, stable orbits* in any solar system. In the RS2 reevaluation, it was found that there is *no preferred direction* of a scalar motion in space, so these eight units are equally divided (±4) about the center of the explosion, which forms a neutral speed zone. So these eight units actually form nine orbits; four inner, a "neutral zone," and four outer, corresponding to the inner planets, the asteroid belt, and the outer planets.

This is an important consequence, in that *all* solar systems are going to look just like ours does. The sizes of planets may be different, and there may or may not be planets in specific orbits, but overall, you'll have none-to-four hard, inner planets, an asteroid belt, and none-to-four gaseous, outer planets. Like our solar system, anything beyond that (Pluto, Charon, Eris, etc) are in *unstable* orbits, primarily being determined by solar gravity and not by their inner cores.

---

152 Larson, Dewey B., _Universe of Motion_, p. 97, where Larson derives the Titius-Bode Law as a natural consequence of the quantum speeds of FTL motion of planets.

Most people are familiar with Zecharia Sitchin's series, the *Earth Chronicles* and the Sumerian cylinder seal VA 243 that he claims to be a depiction of our solar system with an extra planet, *Nibiru*.

However, we know that most solar systems look exactly like our own, so "what if" the solar system depicted on this seal is not *ours*, but *another* solar system? Perhaps the solar system the Annunaki originally came from? December 21, 2012 has come and gone, and no *Cylinder Seal VA 243* Nibiru on our skies… consider the possibilities.

Our early sun is a reconstituted, 3rd generation red giant, a large, relatively cool star with low gravity. The planets, being the shattered remains of its former, white dwarf companion, are still strongly displaced in time with the inner cores providing a significant anti-gravity propulsion system to hold the planets into stable orbits fairly far out, but with each planet moving at a substantially faster orbital velocity[153] then we now observe.

The beginnings of a 3rd generation solar system will initially be a large ring of dust and debris around a dull, red giant sun, with the fragments of the white dwarf maintaining an orbital position in that debris field.[154] Over time, gravity will do its job and the bulk of the debris field will be accumulated by the newly forming sun and planetary cores, making the sun smaller, brighter and hotter, moving it towards the main sequence. The planetary cores cool and expand, with a *Red Giant Sun of Early Planetary System* slowly-increasing layer of rock accumulating over them. The aggregated crust tends to be small, as the anti-gravity motion of the early core will tend to

---

153 The early planetary cores are similar to pulsars in that they have predominant ultra-high speeds in the inner core, producing strong anti-gravity—and anti-mass—effects that virtually neutralize the intrinsic mass of the fragment. As a result of very low net mass, the corresponding orbital velocities are very high, making for short years. Over time, inner core (3-x) degenerates to outer core (2-x) and outer core to mantle (1-x), increasing the mass of the planet and slowing its orbital velocity, making for longer years.

154 This has been observed in protoplanets, the only difference being is that the Reciprocal System places the star in the 3rd generation, not the 1st generation as a "newly forming star," as conventional astronomy indicates.

push away the larger fragments that would otherwise substantially add to the mass. If a space-faring species were to visit the Earth in those early days, after most of the post-supernova debris had been cleared away, they would find eight planets in fairly close orbits, close enough that the rings of Saturn[155] could be observed with the naked eye on near orbital approach, and Neptune would be visible in the night sky much as Jupiter is now.

One of the beautiful parts of Dewey Larson's *Reciprocal System of theory*, is that *everything* works the same way; there is one set of rules that define structure from the smallest electron to the largest supergalaxy, so anything you learn in one, specific field of study is applicable to *all* others. And it is simply based on time and space, the yin-yang of the Universe. If we apply this basic yin-yang knowledge to the newly forming 3rd generation solar system, we find that there are "yin" planets and "yang" planets, distributed around a neutral boundary. Looking at our solar system, the structure becomes obvious: this neutral boundary is the asteroid belt, with the small, condensed "yin" planets being the inner planets and the large, expansive "yang" planets being the outer gas giants. All the inner planets will have similar properties; all the outer planets will have similar properties; and the inner and outer will be conjugates of each other.

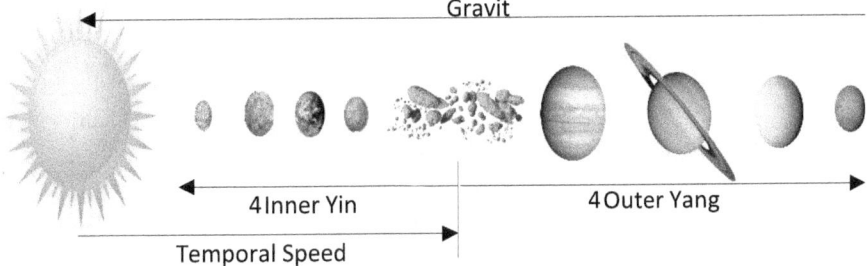

There are a few other consequences of this structure:

1.　　The asteroid belt was *never* a planet, it was a white dwarf star! It exists in a region where forces are relatively balanced, like a stagnant pool. Rock accumulates there and forms asteroids, so over time, a planet-sized asteroid *may* form from gravitation, but that planet will not have a planetary core like the other worlds, and as such, would never be able to sustain an ecosystem.[156]

2.　　The early planets had *no moons*. It is a reasonable conclusion, as during the post-supernova aggregation phase, any moons close enough

---

155 The development of planetary rings occurs during the cooling down stage of the planetary core, **not** being formed *with* the planet. The outer planets contain the larger cores, sufficient to exhibit nova-type explosions that result in many small moons (ultra-high speeds) and ring systems (intermediate speeds).

156 Ceres, now occupying that location, has a rocky core and icy mantle, not an active, planetary core.

to a planet would be sucked in and add to the planet's mass. The moons are a later stage of solar system formation, a product of outer planet nova activity, or someone dropping them off.

3.　　　All planets have a *scalar* magnetic field. Magnetism is a consequence of intermediate speed motion in the Reciprocal System, and all planetary cores have intermediate speed motion. However, without something to give the magnetic field an orientation, the field is random and barely detectable by instrumentality. Once oriented, it normally takes the shape of the classic toroid.

4.　　　The positioning of the planets is arranged about the neutral, unit speed boundary, the asteroid belt—not the sun! The sun controls where this neutral ring sits, in relation to itself. The planets adjust relative to the neutral ring. Gravity is still the controlling factor, but there are *two kinds of gravity*: spatial and temporal (gravity in time; anti-gravity in space).

There are a large number of factors that go into the positioning of planets, not just gravity. The interaction of the various speed ranges of the planetary core, the electric, magnetic and gravitational effects of the sun, and the effects the planets have on each other, since they share a common, white dwarf core that is still localized in time and distributed across space as planets.[157]

### Sister Worlds: the Inner Planets

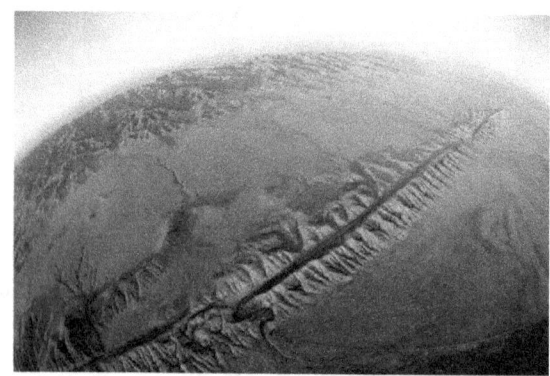

"Yin is in." Yin planets are compressive, resulting in hard, rocky surfaces that crack over time, as their cores cool and expand. This is observed as tectonic plates separating continents on Earth, and rills and fracture zones on the other hard planets and moons. These fracture zones and areas of upheaval are indications that there is a white dwarf fragment at the core of the structure, that is *gravitating in time* and due to the reciprocal relation, expanding in space.

However, without an *oriented* magnetic field, the early *The San Andreas Fault* planets would be subject to exposure of all sorts of ionizing radiation and mass ejecta from the sun, which tends to inhibit, not stop, the development of life. Early forms of life on these worlds will either be resistant to radiation or make use of it as a kind of food source, similar to plants "eating" light through photosynthesis. Once a situation develops where a magnetic field becomes oriented, the resulting strong magnetic field deflects these damaging effects from the sun and an "explosion of life" occurs on the planet, as what

---

157 This temporal locality gives rise to *metaphysical* effects, such as those documented in *Astrology*.

occurred on Earth during the Cambrian period of the Paleozoic Era. This situation occurs when the planet *obtains a moon.*

Initially, the inner planets were in a gravitational lock with the sun. Like our present-day moon, the same face was always pointed at the sun, providing a "light side" and a "dark side."[158] Even now, Mercury and Venus barely rotate on their axes; Mercury's day is longer than its year.[159] This results in a dichotomy of heat and cold, with the face towards the sun getting exceptionally hot. The dark, shadow side, due to little to no atmosphere, will be almost as cold as the surrounding void of space. The thermocline, where hot meets cold, tends to be an area of stress and shear, cracking the young planet around its girth in great and yawning voids… Consider this Norse creation myth:

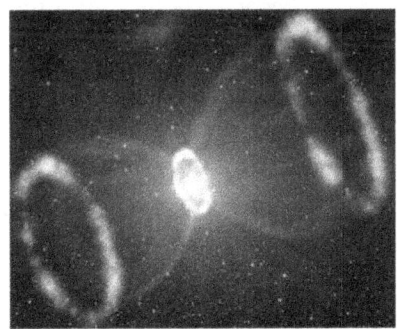

*The first realm to exist was Muspell, a place of light and heat whose flames are so hot that those who are not native to that land cannot endure it.*

*Beyond Muspell lay the great and yawning void named Ginnungagap, and beyond Ginnungagap lay the dark, cold realm of Niflheim.*

Consider it to be a description of the early, inner planets, with the burning-hot side facing the red giant sun, a temperate region around the edges where the sun is low in the sky and the eternal, black cold and darkness of the side facing away from the sun. Again, the yang-boundary-yin structure, with the Norse equivalents being *Muspell* (fire), *Ginnungagap* (the great gap), *Niflheim* (ice).

The sister worlds of Mercury, Venus, Earth and Mars originally looked very similar: dried out balls of rock and volcanoes, cracking their surfaces as the cores expand, with an inverse density gradient putting the heaviest elements near the surface. Not much good for the formation of life. At least not yet.

The ancient records describe things we could not possibly know about, but the "gods" *did* know, relating to us through mythology and apocrypha.

### Brother Worlds: the Outer Planets

The outer planets, being the larger, more energetic fragments, will initially be dark bodies emitting X-ray and gamma radiation (part of their transition from solar to planetary status). As they cool, they will produce significant amounts of visible light, like miniature suns, starting with the most energetic, the outermost planet of Neptune, and moving in quantized steps inward. In essence, they behave like their larger cousin,

---

158 This dichotomy of light and dark forms the essence of yang and yin; yang being the sun-side, and yin is the shaded.

159 Mercury's year is 87.97 Earth days, whereas one day, sunrise-to-sunrise takes 176 Earth days.

the nova, as the cooling process will produce light elements that gravitate towards their centers, including hydrogen and oxygen—an explosive combination.[160] And eventually

they do go off with a bang, scattering dust, rock and planetary *belt. Planetary rings form the SAMESN 19878A Creating an asteroid* chunks into nearby space, typically along the equatorial plane, which *way and for the SAME reason.* produce a series of icy moons and rings, much like the supernova *Saturn would initially look like this,* that created the planets and asteroid belt. Again, it is the *same with bright, visible, glowing rings. process,* just a different scale.

After their blaze of a nova subsides,[161] they will have large atmospheres of light elements, with a ring system and a number of moons scattered about. If the explosion is energetic enough, some of these moons may reach escape velocity and go wandering to the outer solar system (anti-gravity motion), to take up unstable orbits further out.[162]

The sequence of events in the ancient sky, because of quantized transition and remembering that all the planets *share* the same core in 3D time, will be that when one planet drops in energy, the next lights up, a kind of "stair-step" effect.

As a result, there will be a dominance of brilliant, star-like planets in the sky:

- At first, no visible planets, but point-source X-ray and gamma ray sources are detectable where the planets will "form" (they are already there, just not emitting nor reflecting light).

- Neptune flares; the other planets are still dark, X-ray bodies. At this time, the planetary orbits of both inner and outer planets are much closer to the unit speed boundary, the asteroid belt, and are just now beginning to spread away from it.

- Neptune dies down to a planet with a ring system; the lower speeds move the planet into a more distant orbit. Uranus flares up and becomes dominant in the night sky, with Jupiter and Saturn still being dark bodies.

- Uranus dies down from the same conditions and Saturn

---

160 Larson, Dewey B., *Universe of Motion*, Chapter 13, "The Cataclysmic Variables," p. 182. The planetary "nova" is the same explosive process as the stellar nova, which is timed by the change in magnetic ionization levels.

161 The nova flare may last for months, due to the smaller size of the core as compared to the white dwarf star. The higher luminosity can remain for centuries, until sufficient material drops below the FTL speed range to darken the disk.

162 Pluto, Haumea, Makemake and Eris are likely candidates for this situation.

becomes active, leaving only Jupiter a dark body.

- Saturn dies down and Jupiter flares up, becoming the dominant object in the sky.

- Jupiter dies down to planetary status. By this time, the planets have lost most of their antigravity motion, their orbital velocities have reduced, and they are in orbits much further apart from each other, but also will have moved closer to the sun. The sun has been condensing all along and getting smaller and brighter, heading towards the main sequence dwarf.

This is the current structure of the solar system. It has change *significantly* since its formation.

Note that the sequence of bright, star-like objects in the sky, a natural consequence of Reciprocal System astronomy, matches the mythological Ages (or "suns" as the Aztec say) or the domination of certain Titans: Poseidon (Neptune), Ouranos (Uranus), Cronus (Saturn) and Zeus (Jupiter). The changing dominance of these very obvious planetary "lights" in the sky plays a significant role in mythology, as the planetary orbits also change when the "ruler of the sky" changes. For example, when Jupiter became dominant over Saturn in the sky, the orbit of Saturn moved further out and the ring system could no longer be seen— Zeus/Jupiter overthrew his father, Cronus/Saturn, whom overthrew his father, Ouranos/Uranus. There is a mythological pattern that mimics the consequences of Reciprocal System astronomy.

The inner planets have not changed much, other than to slow down and move closer to the sun, as the core sizes are too small to sustain the drastic exhibitions of the outer planets. The occasional core flare will crack their surfaces and cause sudden expansion.[163]

Magnetic Ionization of Elements

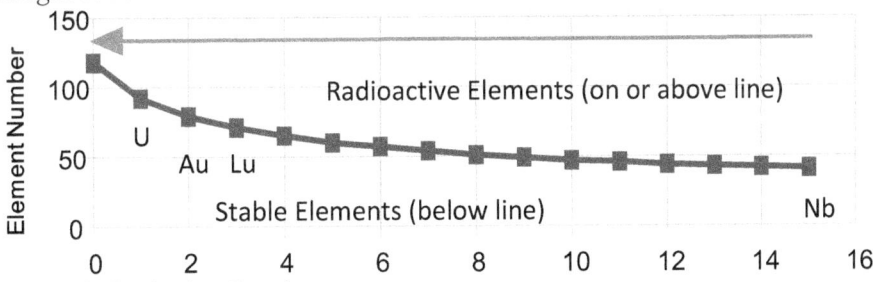

Magnetic Ionization Level

Magnetic ionization levels were initially very high because the core is a stellar fragment and decreases as the planetary core ages. Rather than a solar transition up to a higher magnetic ionization, the planets do a "planetary transition" down to a lower magnetic ionization. This infers that much of the material on the early planets was *radioactive*, and as the planets age, their minerals become less

---

163 Peret, Bruce, "At the Earth's Core: the Geophysics of Planetary Evolution," discussion of "core flares," a volcanic-like eruption that later looks like a meteor impact crater.

radioactive.

The chart indicates the zone of stability for elements at a given magnetic ionization level (using Larson's "natural units" of magnetism). All the elements below the line will be stable in the environment; all the elements above will be radioactive. For an example, at a magnetic ionization level of 2, the stable elements are from hydrogen (1) to platinum (78). The radioactive elements would start with Gold (79, Au) on up to element #117, which is the maximum element[164] in the Reciprocal System.

One of the major consequences of the planetary cores having the similar inverse density gradient of the white dwarfs, is that the *heaviest elements are on the outside*, and the lightest ones are deep underground, until meteoric aggregation covers them up with dirt, and they have to be dug out.

Consider: if a bunch of extra-terrestrial engineers were out to exploit some "heavy metal," along with all the gems produced by extremely high surface temperatures of the early core, the inner planets of a newly forming solar system would literally be a "gold mine."

### Using *Atlantis* to Correct Geochronology

People have suspected for a long time that our world has been visited by extraterrestrial species, as there is a considerable amount of evidence laying about that would indicate visitation by a space-faring race.[165] The difficulty has been that researchers could not correlate the evidence with a mythological chronology or geography, because of the bad assumptions concerning geologic dating and the belief that the Earth always looked just as it does now. When we corrected the interpretation of years for the

*City of Poseida* Mayan calendar, it matched the Hebrew calendar. The same thing *on the* happens when the "billions and billions of years"[166] of *Continent of Atlantis* geochronology becomes "thousands and thousands."

---

164 Element 117, "Larsonium," as Reciprocal System students call it, is the upper limit of rotational combinations that can be expressed in a 3-dimensional system. This does not preclude elements with a higher number, but there is no way to express those combinations in 3D space, so elements are highly unstable, both structurally and chemically. They normally decompose to a lower element within one natural unit of time (approximately 152 attoseconds).

165 Von Däniken, Erich, *Chariots of the Gods*.

166 A popular phrase used by astronomer Carl Sagan.

There are a number of sources that can be used to build a correlation between the accepted geochronology and the corrected calendar; since Edgar Cayce's references to *Atlantis* are fairly well known, I will start with them. Cayce identified three "destructions" of Atlantis:

1.  50,000 BCE[167]: a technological attempt to eliminate the giant beasts ravaging the land failed, and resulted in a major upheaval of the land splitting the continent into the three, large islands of Poseid, Aryan and Og.

2.  28,000 BC: earthquakes and flooding that resulted in Poseidia's climate changing for the worse, to the point where the island was evacuated before it froze over.

3.  10,000 BC: the final breakup of the islands of Atlantis.

When a person receives information from a non-sensory source, like channels or dreams, the expression of the information is limited to the symbols and motifs that the person is familiar with. This has been scientifically demonstrated in situations where primitive tribes describe aircraft as giant birds. The same situation works with chronology; when Cayce selected these dates, they were based on the information he accepted as truth, the geochronology of the period. In the case of the final breakup of

Atlantis, the date was commonly accepted as 9600 BCE, which comes from Plato's recounting of Solon's story of the continent, 9000 years before his time. Solon's time was circa 600 BCE, so 600+9000 = 9600, the accepted time of the fall of Atlantis that was known to Cayce.

Then along comes one Angelos Galanopoulos, pointing out in his 1969 book, *Atlantis: The Truth Behind the Legend*, that there was a slight error in the Egyptian translation of 9000 years—it was actually 900 years, making the time frame of the destruction of Atlantis about 1500 BCE (600+900). From the prior exploration of the calendar, we know that the end of the 4th Baktun was 1548 BCE, corresponding with the Exodus from Egypt, the plagues, winds, shaking of land and fire falling from the sky. The final destruction of Atlantis now corresponds with the Mayan and Hebrew records of the period, and gives us a working point to make an educated guess at the 1st and 2nd destructions.

We already know the correlation is off by approximately 10:1, we can estimate that the 1st destruction of Atlantis was no more than 5000 BCE, and the 2nd no more than 2800 BCE, based on Cayce's information. We also know that the year counts are off due to the "number of days per year" changing, we can make a

---

167 Cayce reading 262-39; "50,722" BCE, referring to the time of a gathering, "…except in the inner thought or visions of those that have returned or are returning in the present sphere, the ways and means devised were as those that would *alter or change the ENVIRONS for which those beasts were needed*, or that necessary for their sustenance in the particular portions of the sphere, or earth, that they occupied at the time." These "returned or are returning" souls are probably the ones behind the chemtrail geoengineering.

quick *estimate* of what year 5000 BCE is actually referring to, on the corrected calendar. The early calendar was 260 days/year, and the final switch to 365 took place in 749 BC. So if we take the pre-749 BC years and adjust, (5000-749) x 260 / 365 = 3026 years past 749 BC, making the 5000 BCE date on our corrected calendar to be *around* 3775 BCE. And right there, at 3761 BCE, we have the start of the Mayan and Hebrew calendars, defining the creation of man.

If we apply the same logic to the 2800 BCE date, we end up at 2209 BCE, corrected calendar, with the Biblical Flood sitting at 2105 BC, a very close match. So we can now draw the following correlations as a starting point:

| Year, BCE | Year, AM | Long Count | Accepted Year, BCE | Events |
|---|---|---|---|---|
| 3761 | 0 | 0.0.0.0.1 | 50,000 | Biblical Creation of Adam & Eve. Mayan Creation of Human World. First destruction of Atlantis. |
| 2105 | 1656 | 2.19.16.0.0 | 28,000 | Biblical Deluge. Mayan Deluge, only about 5.5 modern years from the end of the 3rd Baktun. Second destruction of Atlantis. |
| 1548 | 2213 | 4.0.0.0.0 | 9,600 | Hebrew Exodus from Egypt. Third destruction of Atlantis. |

The inference is obvious; mankind was created by "God" or "gods" around 3761 BC, which we are incorrectly dating as 50,000 BCE. So it may be interesting to see what else may have been going on at

50,000 BCE that might be of interest, and we find: "These 'types' of Early Modern Humans

*Using Atlantis to Correct Geochronology*
[Neanderthals] supposedly evolved into the next step toward actual humans—the Cro-Magnons, living

50,000 to 60,000 years ago."[168] With the corrected calendar, 50,000 BCE becomes 3761 BC and the

Cro-Magnons, **us**, showed up in the geological record at the *same time* as Adam and Eve and the Mayan creation of man, right after the first destruction of Atlantis, and **missing a link** to the Neanderthals that were previously overrunning the land, much to the annoyance of the Atlanteans.

This is congruent with the Biblical accounts of the creation of Adam and Eve being the first *humans* that God created on Earth. First humans, yes, but *not* the first intelligent species to exist on Earth. The Atlanteans were intelligent—*just not human*—so they predate the Biblical accounts.

---

168 Pye, Lloyd, *Intervention Theory*, p 153.

Of course it also begs the question of "Who is this God-Person, Anyway?"[169] showing up in mythology all over the planet, creating humans? This will be the subject in another paper of this series, but based on my Montauk experiences, I can sum it up in 3 letters: SMs. The Saurians known to the Sumerians as the *Annunaki*, to the Christians as the *Elohim*, to the Norsemen as the *Æsir*, to the Indians as the *Asura*, and by many other names, all over the globe.

**Planetary Moons**

As previously mentioned, the early planets had no moons because all the fragments of the white dwarf destroyed in the supernova explosion would have predominant, ultra-high speed motion—anti-gravity —and therefore would move away from each other. The outer planets would experience nova-like explosions over the course of their cooling down, ejecting both planetary rings and many smaller moons. But what of the inner planets? Mercury and Venus have no moons, Earth has an exceptionally large one, for the size of the planet, and Mars has a couple of very small moons that don't behave properly, orbiting in the wrong direction and faster than the planetary rotation.

Immanuel Velikovsky had some interesting perspectives with his rogue planet theories, but the difficulty with them is that planets, containing white dwarf cores are subject to the properties thereof, are not able to leave their orbital positions and go wandering about the solar system. Even the moonsized ejecta from the outer planets would be unlikely to get past the neutral zone of the asteroid belt to reach the inner worlds. Asteroids and comets, containing no intermediate or ultra-high speed matter, operate solely on gravitational principles and can drift all over the solar system, but because they lack those very motions, they can never form a stable orbit around a planet.

The moons of Earth and Mars have no apparent, *natural* origin. In *The Case for the UFO*, Jessup writes:

"Dean Swift was prescient in regard to his astronomy, predicting that Mars had two small satellites, one of which was close to Mars' surface and made two revolutions daily. It has been pointed out that this inner body is too close to Mars to be in adjustment with any known postulate of the natural distribution of satellites relative to their parent body. This may be an indication that Mars' inner satellite is artificial."[56]

In the Varo annotations of the book, it is noted that the Martian moons are "Also an old 'Dead-Ark' S-
M MAKE."

*Noah! I Want You to Build... an Ark! Right! What's an Ark?*[170]

---

169 Adams, Douglas, *The Hitchhikers Guide to the Galaxy*. The final book of Oolon Colluphid's trilogy, "Where God Went Wrong," "Some More of God's Greatest Mistakes," and "Who is the God-Person, Anyway?" 56 Morris, Jessup K., *The Case for the UFO*.

170 Cosby, Bill, *Bill Cosby is a Very Funny Fellow: Right!*, Warner Brothers Records, 1963, "Noah."

An "Ark" is a term used by both the LMs and SMs to refer to the large, interplanetary "motherships" that they use as colonies and transportation between solar systems. They are constructed from asteroids that have intermediate speed motion (small, planetary "cores") and therefore an *inverse density gradient* to their structure. With all the heavy, hard metals on the outside providing protection from the ravages of space and a "soft, chewy center,"[171] of lighter elements and atmospheric gases at the core, they are essentially a "prefab spaceship" ready for interstellar travel; being a smaller application of a Dyson shell.[172] All that is required is a navigation system to orient the scalar magnetic field, already present from the intermediate speed motion, into either *paramagnetic* (attractive) or *diamagnetic* (repulsive) modes.

Also noteworthy in these Ark designs is that if ultra-high, anti-gravity motion is still present, as it is in planetary cores, the center of the atmospheric cavity will contain a small, sun-like object representing that "inner core" construct; a structure that exactly matches the "hollow Earth" theories. When the intermediate speed motion of the Ark finally slows to sub-light speeds, the ark "dies" and is no longer functional as a spacecraft. This is what the Varo annotations are referring to concerning the Martian moons as "Dead Arks." (Montauk had references of there being a large number of "dead arks" scattered throughout the solar system, as they don't last forever and are abandoned.)

If the Moon were an Ark, then there should be references to times when it was not in orbit, or missing altogether, in mythology. In his paper, "The Earth Without the Moon,"[60] Velikovsky cites many sources of a time when there was no moon in the sky:

The period when the Earth was Moonless is probably the most remote recollection of mankind. Democritus and Anaxagoras taught that there was a time when the Earth was without the Moon.[61] Aristotle wrote that Arcadia in Greece, before being inhabited by the Hellenes, had a population of Pelasgians, and that these aborigines occupied the land already before there was a moon in the sky above the Earth; for this reason they were called Proselenes.[173]

---

171 An old, commercial description of a "Tootsie Pop" with a hard, candy shell and a soft, chewy center that depicts the inverse density structure.

172 A spherical shell surrounding a sun, to capture all of its energy output, incorrectly called a Dyson Sphere in the *Star Trek: The Next Generation* episode, "Relics," where they found Mr. Scott's shuttlecraft crashed on its surface. 60 Sammer, Jan, *The Velikovsky Archives*, http://www.varchive.org/itb/sansmoon.htm 61 Hippolytus, *Refutatio Omnium Haeresium* V. ii.

173 Aristotle, fr. 591 (ed. V. Rose [Teubner:Tuebingen, 1886] ). Cf. *Pauly's Realencyclopaedie der classischen*

*Altertumswissenschaft*, article "Mond" ; H. Roscher, *Lexicon d. griech. und roemisch. Mythologie*, article "Proselenes." 63 *Argonautica* IV. 264.

Apollonius of Rhodes mentioned the time "when not all the orbs were yet in the heavens, before the Danai and Deukalion races came into existence, and only the Arcadians lived, of whom it is said that they dwelt on mountains and fed on acorns, before there was a moon."[63]

Plutarch wrote in *The Roman Questions*: "There were Arcadians of Evander's following, the so-called pre-Lunar people."[174] Similarly wrote Ovid: "The Arcadians are said to have possessed their land before the birth of Jove, and the folk is older than the Moon."[175]

*Planetary Moons*

Hippolytus refers to a legend that "Arcadia brought forth Pelasgus, of greater antiquity than the moon."[176] Lucian in his *Astrology* says that "the Arcadians affirm in their folly that they are older than the moon."[177]

Censorinus also alludes to the time in the past when there was no moon in the sky.[178]

Some allusions to the time before there was a Moon may be found also in the Scriptures. In Job 25:5 the grandeur of the Lord who "Makes peace in the heights" is praised and the time is mentioned "before [there was] a moon and it did not shine." Also in Psalm 72:5 it is said:

"Thou wast feared since [the time of] the sun and before [the time of] the moon, a generation of generations." A "generation of generations" means a very long time. Of course, it is of no use to counter this psalm with the myth of the first chapter of Genesis, a tale brought down from exotic and later sources.

The memory of a world without a moon lives in oral tradition among the Indians. The Indians of the Bogota highlands in the eastern Cordilleras of Colombia relate some of their tribal reminiscences to the time before there was a moon. "In the earliest times, when the moon was not yet in the heavens," say the tribesmen of Chibchas.[179]

Religious apocrypha also relates that during the time of Adam in the Garden of Eden,[180] the sun remained *fixed* in the eastern sky and the cycle of days and nights

---

174 Plutarch, *Moralia*, transl. by F. C. Babbit, sect. 76.

175 *Fasti*, transl. by Sir J. Frazer, II. 290.

176 *Refutatio Omnium Haeresium* V. ii.

177 Lucian, *Astrology*, transl. by A. M. Harmon (1936), p. 367, par. 26.

178 *Liber de die natali* 19; also scholium on Aristophanes' *Clouds,* line 398.

179 A. von Humboldt, *Vues des Cordillères* (1816), English transl.: *Researches Concerning the Institutions and Monuments of the Ancient Inhabitants of America,* (1814), vol. I, p. 87; cf. H. Fischer, *In mondener Welt* (1930), p. 145.

180 "Garden" is derived from "guarded," a *protected enclosure*. Eden is E-

only began upon their expulsion from the garden, which we now know correlates to the time of the first destruction of Atlantis—something big changed. (They *did* measure a daily cycle—the movement of the "celestial chariot" across the sky, as is common in many ancient myths. "Seven times the Lord crossed the heavens," the seven *domas*, or *hebdomas*, what we now call a "week," was original a measurement of a day, when the Earth had none.

But the descriptions refer to a "Lord" meaning "shining star," not something the size of the moon. The Garden had no nights, but did have an orbital object bright enough to be visible in the light of day.)

The observations that the Earth did not have a moon, and did not rotate on its axis (tidal lock with the sun), are *supported* by the natural consequences of Reciprocal System astronomy. Again, we find science, religion and myth all saying the same thing.

I did some checking of Velikovsky's references and found that the Moon actually *appeared* and *disappeared* at fairly regular intervals during pre-Adamic times, the *divine year*. Velikovsky attributes this to a wandering Moon intersecting the orbit of Earth, until it was finally captured. It sounds good, but for one "reciprocal" exception. Moons and planets, *having* motion in time, will behave in the same manner as *atoms*, having their motion in time. When atoms go into "stable orbit," we call it *chemistry*. That chemistry is based on the concepts of *valence*,[71] which is just "matching speeds" in the Reciprocal System. The probability against a planet and moon to having exactly the right "chemistry" to achieve stable orbit is, shall I say, astronomical?

The largest factor is overcoming the *unit speed boundary*, the same limit that keeps all the atoms in the Universe from coming together to form a single, super-molecule of everything. The only way a moon could establish orbit around a planet, given the chemical requirements of the Reciprocal System, would be to adjust its motions as it approached to match the new environment. Not likely for a random fragment of a white dwarf, but *very* likely for an Ark, under intelligent control.

The natural consequences of our theory indicate that the moons of the inner planets are not "natural," in the sense of evolving with the associated planets, but are actually "Arks" that were used by the SM Annunaki, placed in orbit around Earth and Mars, not very long ago. The Martian Arks of Phobos and Deimos are "dead," and are now just asteroids in decaying orbit. However, the Earth's Ark, the Moon, is still running on impulse power with its inverse density gradient, hard shell and gaseous core.

Consider NASA's "lunar mysteries" in this light:[181]

---

DIN, the Sumerian "abode of the righteous ones." 71 Larson, Dewey B., *Nothing But Motion*, Chapter 18, "Simple Compounds."

181 Childress, David Hatcher, *The Anti-Gravity Handbook*, "Eleven Things That NASA Discovered About The Moon That You Never Knew." (Summary of points.)

1.      Scientists now tend to lean toward the third theory—that the moon was "captured" by the earth's gravitational field and locked into orbit ages ago. Opponents of the theory point to the immensely difficult celestial mechanics involved in such a capture. All of the theories are in doubt, and none satisfactory. NASA scientist Dr. Robin Brett sums it up best: "It seems much easier to explain the nonexistence of the moon than its existence." [Captured, or parked?]

2.      Incredibly, over 99 percent of the moon rocks brought back turned out upon analysis to be older than 90 percent of the oldest rocks that can be found on earth. [The Annunaki are not from *this* solar system; their Arks would be much older than the planets.]

3.      The mystery of the age of the Moon is even more perplexing when rocks taken from the Sea of Tranquility were young compared to the soil on which they rested. [Meteoric aggregation, just like the crust of the planet over the mantle, is much younger.]

4.      During the Apollo Moon missions, ascent stages of lunar modules as well as the spent third stages of rockets crashed on the hard surface of the moon. Each time, these caused the moon, according to NASA, to "ring like a gong or a bell." On one of the Apollo 12 flights, reverberations lasted from nearly an hour to as much as four hours. NASA is reluctant to suggest that the moon may actually be hollow, but can otherwise not explain this strange fact. [Inverse density gradient makes it hollow.]

5.      Astronauts found it extremely difficult to drill into the surface of these dark plain-like areas [maria]. Soil samples were loaded with rare metals and elements like titanium, zirconium, yttrium, and beryllium. This dumbfounded scientists because these elements require tremendous heat, approximately 4,500 degrees Fahrenheit, to melt and fuse with surrounding rock, as it had. [A white dwarf fragment has an initial surface temperature of 180,000° F— definitely hot enough to do that.]

6.      The Soviets announced that pure iron particles brought back by remote controlled lunar probe Zond 20 have not oxidized even after several years on earth. [From Larson's chemistry, $Fe_5$, which cannot oxidize.]

7.      The upper 8 miles of the moon' crust are surprisingly radioactive. [Starts at a high magnetic ionization, and works down, creating many radioactive elements. Though I do wonder how NASA got "8 miles" of core samples from equipment in that tiny lunar module.]

8.      But after Apollo 15, NASA experts were stunned when a cloud of water vapor more than 100 square miles in size was detected on the moon's surface. … The water vapor appears to have come from the moon's interior, according to NASA. [Hollow, gaseous interior leaking out.]

*Planetary Moons*

9.      Lunar explorations have revealed that much of the moon's

surface is covered with a glassy glaze, which indicates that the moon's surface has been scorched by an unknown source of intense heat. [Prior to being an Ark, the lunar surface WAS a source of intense heat.]

10.    Early lunar tests and studies indicated that the moon had little or no magnetic field. Then lunar rocks proved upon analysis to be strongly magnetized. [Intermediate speed motion.]

11.    In 1968, tracking data of the lunar orbiters first indicated that massive concentrations (mascons) existed under the surface of the circular maria. NASA even reported that the gravitational pull caused by them was so pronounced that the spacecraft passing overhead dipped slightly and accelerated when flitting by the circular lunar plains, thus revealing the existence of these hidden structures, whatever they were. Scientists have calculated that they are enormous concentrations of dense, heavy matter centered like a bull's-eye under the circular maria. [Core flare from early expansion activity.]

Considering the Moon as a white dwarf fragment that was converted to an Ark for space travel, readily explains *all* the observed anomalies, within the context of the Reciprocal System. And as a functional Ark, the moon could easily arrive and depart Earth's orbit at the *will* of its operators.

### New Jerusalem

The *Book of Revelations* describes the city of God as *New Jerusalem*, giving a description of its size:

*And the city lieth foursquare, and the length is as large as the breadth: and he measured the city with the reed, twelve thousand furlongs. The length and the breadth and the height of it are equal.*

Revelation 21:16

If this is an astronomical object, its "Borg Cube" shape would have undoubtedly been covered in dust and debris over the centuries, turning it into a sphere and giving it the appearance of a celestial body. So, let us "do the math" and see if we can find something matching the description.

First, we have to determine the size of a *furlong*, in modern measurement. The furlong has changed values a few times, but is (was) accepted as 600 feet, reminiscent of the sexagesimal system used by the Sumerians ($60 \times 10$). But how big was a foot, 2000 years ago, in Mesopotamia? Values on record range from 250mm to 330mm (France), with the accepted value being the British foot of 305mm. The "bigfoots" of Napoleon's era seem to be the exception to the rule and Mesopotamians are physically smaller people than Englanders, so let's split the range difference, $(250+305)/2 = 278$mm, as an estimate of the actual size of a Biblical "foot." Now to some calculations:

$278/305 = 0.91$ of the normal "foot" size.

$0.91 \times 600$ feet $= 547$ feet per furlong (ancient values were estimated around 550 feet). 547 feet $\times 12000 / 5280 = 1243$ miles on a side (not the accepted 1500 miles).

To find the circumscribed sphere that would account for the dust and rock of centuries, we multiply the side by $\sqrt{3}$, giving a sphere that is at least 2153 miles in diameter. Taking a quick look at objects in our solar system, we have a potential "winner" with a mean diameter of 2159 miles, also having some very unusual, physical properties like being hollow—*our Moon*. Could it be that New Jerusalem is already here, parked in orbit around Earth? An "Ark?"

**Pre-Cro-Magnon Geochronology**

Attempting to calculate the geochronology of the period prior to the creation of man and the first destruction of Atlantis is challenging, as the Earth was not rotating on its axis, was physically smaller, in a different orbit, and the basic intervals of measure are drastically different.

The Sumerian *Kings List* is said to document the initial arrival of the Annunaki some 241,200 years ago

—but it does not actually *say* that. What it lists are 8 kings that ruled for a total of 66 *sars*, 6 *ners*.[182]

And the ancient Sumerians use "dates" differently. For example, there are intervals for 1 day, 30 days

(month) and 360 days (year), but there is no "12 months in a year" concept. A month is a "watch" (1/12) of a "year" (360/1) = "month" (30/1). I'm not an "expert," but it would seem that *sars* would be a divine year of 3600 *days*. If we make that assumption, the longest reign of a king on the *Kings List* was 12 *sars*, which would be 43,200 *days* or about 118 modern years. As in most ancient cultures, Kings were Kings since birth. To quote Genesis 6:3, "And the LORD said, My spirit shall not always strive with man, for that he also is flesh: yet his days shall be a *hundred and twenty years*." So King En-men-lu-ana just made it in "under the wire" at 118. That also means that the reign of the Annunaki gods in Atlantis was only about 653 modern years, not 241,200 years, which again supports the premise of this paper that geochronology has been greatly exaggerated, to hide the history in the past.

**Conclusion**

I know I've raked a lot of concepts over the coals in this paper, from geochronological time lines, Atlantean civilization, the formation of solar systems and backwards stellar evolution, showing there was never any planet that made the asteroid belt, disproving Velikovsky's rogue planet theories and pointing out Sitchin may have had the wrong solar system. Not only have I turned science upside down (which is easy, because everything is backwards), but also rocked the boat on both religions and New Age beliefs.

But consider this: now that you have a working understanding of a very advanced, spatio-temporal "theory of everything," the *Reciprocal System*, that corrects errors in modern science, has its roots firmly planted in the ancient traditions of yin-yang and is working in sync with mythology and religion… for the first time you can actually *deduce* what is "out there," as well as know where to look for it. Using Reciprocal System astronomy, we know *exactly* the stellar conditions to look for, for brave, new worlds for us to colonize, or exactly where worlds with extremely

---

182 Sumerian system of temporal measurement; 3600 *sars*, 600 *ners* and 60 *sosses*, which are "periods," not years. The translators just assumed "sar" equaled "year" because of the large value, and that may not be the case. Certainly *not* a 365-day, modern year!

advanced civilizations are likely to be. No more guesswork, channeling, or trying to translate ancient records. Just "natural consequences" arising from the way the Universe was put together.

For centuries, people have wanted to understand the Universe to find their place in it. Well, now you have a map, complete with directions, rest stops and tourist highlights. We can continue to stick our heads in the sand with wild suppositions, hopes and dreams, or just grab on to this new understanding, learn and teach our brothers and sisters, and become that "good neighbor" to take our place in the Universe and pursue our own destiny.

# NEW WORLD RELIGION
# ENSLAVING THE HUMAN SPIRIT WITH A BLUE BEAM

**Preface**

This is a discussion of *Project Blue Beam*, a plan that was created some 50 years ago by the folks bringing you a "new world order" at the onset of the hippie/New Age movement, to address the developing spirituality in *homo sapiens*—and what could be done as a long-term solution to control and/or eliminate it, along with the sense of freedom and individuality that the spirit complex tends to bring out.

My original understanding of *Blue Beam* was that it was a holographic technology that was bought out by the military-industrial complex from some New England inventors back in the mid-1970s. I was in High School at the time and recall a short news report on it, then never heard anything about it again. It was only recently that I ran across Serge Monast's exposé[5] on *Project Blue Beam*, and when I put 1+1 together and got 10,[6] I realized that I had seen a lot of the early tech that was destined for this project. So I did some checking on recent developments and was quite astounded at the technological developments that I found... got to tip my hat to those NWO scientists, as it is some brilliant work and they have really taken it a long way down a difficult path. It's just a shame that all that effort is going in to destroy the human spirit, rather than to encourage it.

**Background**

Some of the concepts of my original "Part 2" are needed to understand what is going on with *Blue Beam*. A quick summary is that our world was colonized by a saurian race (whom "those in the know" refer to as "SMs") just as native life was developing. Attempts to turn the natives into a slave force failed, and these saurians used their genetic engineering skills to create a hybrid slave force by combining the "good slaves" they brought with them (a kind of automaton, that were biologically unsuitable to the early, radioactive planetary environment), cross-bred with the local apes and Neanderthals, resulting in Cro-Magnon man. These stories formed the "creation of man" basis of many of the theologies we have today, which, of course, have been revised many times over the centuries to make them more acceptable, as mankind evolves and develops.

One thing that mankind has always been very skilled at is *historical revisionism*. That same revision process continues, under *Blue Beam*, where the old gods are now being supplanted by *ETs* (extraterrestrials) or the new fad, *EDs* (extra-dimensionals), to meet the demands of a more scientific and technical community. Out with the "chariots of fire" and in with the anti-gravity starships!

## The Human Role: Slavery

One of the obvious conclusions of this "hidden history" of mankind is that Cro-Magnon man was created for a *specific* purpose: to be *slaves to the gods* and *controlled* by their direct descendants, now known as the "New World Order."[183] This modality of thought is ingrained in the genetic code of human biology, so it is difficult to *consciously* comprehend—let alone, consciously *override*. Even today, most people spend their lives looking for someone to follow, worship, or obey—whether it be a religious figure, new age guru, government bureaucrat, successful corporate leader, or movie star. Like a cog in a big machine, that's the role you were genetically engineered to fulfill: *slavery*.

The original colonists of our world were the race referred to in mythology as the *Titans*. Their descendants, known in Sumerian as the *Annunaki* (the Christian *Elohim*), became the basis for our classical gods—and the religious belief system *does not matter*, as they are *all* referring to the same group of SMs. What many people do not realize is that with some 5773 years of "royalty" in charge of the planet, they have managed to get control of *all aspects* of society. The 1978 British Sci-Fi series, *Blakes 7*, makes a good point of this in the episode, "Shadow" (an addictive drug), where Blake and his freedom fighters attempt to use the mafia-like *Terra Nostra* against a corrupt *Federation*, with a surprising conclusion:

Avon: This is the I.D. of a guard I killed. He was a member of Federation security—a very special member. He was one of the President's *personal* security force.

Blake: The President of the Federation *runs* the "shadow" operation. Avon: And since "shadow" is the basis of the *Terra Nostra*… Gan:      I don't believe it!

Blake: It's quite logical. To have *total control*, you must *control, totally*. Both sides of the law. The *Terra Nostra;* the *Federation*: two sides of the *same* power. The *same man of power*.

Avon: Ironic, isn't it? We were hoping to use the *Terra Nostra* to attack the Federation, only to discover that it is already being used to support it.

        Vila:      Where are all the good guys?

Blake: You could be looking at them.

And we have the same situation, here: two aspects of the same power: *political* leaders and *religious* leaders. Two sides of the same, new world power. Whether you realize it or not, we are *already* operating under a "one world government," as well as a "one world religion," separated by artificial boundaries called *countries* and *faiths*, to be "divided and conquered" by a group that believes you should murder one another, simply because "they" were born on *that* patch of dirt, while "you" were

*The Human Role: Slavery*

---

183 The "New World Order" consists of the "royal" lines of kings, queens, presidents, multi-national corporate leaders, popes, gurus, priests, llamas, lords and ladies that run the governments *and* religions of the worlds. Also known as the "Cabal."

born on *this* patch of dirt. Or perhaps you believe the Donut Diner employee's hats should be blue, while the "enemy" believes they should be red.[8] They use *any* excuse for conflict, for without all these artificial "lines in the sand," we would probably get along just fine without their dictates.

One of the premises of faith is that "god," with whatever label you stick on him/her/it, is all-powerful, all-seeing and all-knowing... so what do gods need with slaves? Unless, perhaps, they are not as allpowerful as we have been led to believe.

By using the "theory of everything" concepts of Dewey Larson's *Reciprocal System* to get our planetary evolution data in the right direction, mythology and legend tell a slightly different story, one where the "gods" were an advanced, extra-terrestrial race that came here to exploit the riches of a newly forming planet. Based on the the minerals and materials collected, and remnants of fused, "green glass" indicating the use or testing of nuclear weapons around the world, they were probably *arms dealers*, not spiritual leaders. And when one examines mythology from that premise, there is a good deal of supporting evidence. The "like father, like son" or "created in God's image" behavior of humanity is strong evidence of that.

So why the need for slaves? The ancient gods were about as likely to go out and dig up radioactive elements with their hands as Hilary Clinton would be to sponsor a gun show. So they needed a large, *expendable* group to do their "dirty work." According to Sumerian legends, the automatons they brought with them from prior expeditions did not do well here on Earth and expired quickly— maintenance was just too high for good productivity. Hence the need for a more genetically-compatible slave force, better equipped to do the work in this environment. And the "locals," the Neanderthals ("Yeti" or "Sasquatch" as we would call them today[9]), did not seem to have much difficulty with the harsh, radioactive environment of early Earth. Sounded like a good mix for a hybrid; the slave mentality of the SM automatons, plus the sturdy genetics and robust bodies of the local inhabitants.

The early attempts were minor genetic alterations of the SM automatons, as the gods wanted to keep the slave mentality fully intact. But these original hybrid slaves turned out to be idiots, and they kept spilling the elixirs and tripping over the ottoman when entering a room[10]—and something had to be done about that. Enki, the Annunaki "science officer" and Ninhursag, the "chief medical officer," came up with a plan to make these hybrids more intelligent, so they would be better slaves and able to anticipate the needs of the gods and fulfill them. But not *too* intelligent—that was reserved for the children of the gods, only, along with those tall hats they like to wear to either hide, or emulate, those extended craniums.[11]

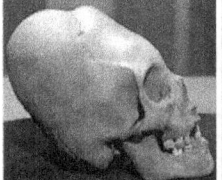    

8    From the British Sci-Fi series, *Red Dwarf.* The race of *felis sapiens* almost fought themselves to extinction over this issue, when in reality, the hats were supposed to be green.

9    Pye, Lloyd, *Intervention Theory*, discussing the physical similarities between Neanderthal and the Yeti/Sasquatch creatures, postulating that Neanderthal continues on today *as* these species. A position I completely agree with.

10    Dick Van Dyke, *The Dick Van Dyke Show*, introduction, where the character Rob Petrie pratfalls over the ottoman.

11    The Annunaki descendants had large skulls, extending up and back, over which they wore hats and other decorations. People of "importance" emulate this behavior, such as religious leaders, wearing hats that extend up in a dome shape behind them. The extended cranium was an outward sign of being a descendant of a god.

Cloning was an unsure and time-consuming process, and Enki was in charge of that, so he got the idea to go behind the back of his brother, Enlil (General of the Annunaki military, always demanding more slaves) and did some unauthorized "updates." Enki introduced genetic modifications to the functional saurian-Neanderthal hybrid slaves in the laboratory cage, the guarded enclosure of E-DIN[184], by introducing a retrovirus to make their test subjects more mammal-like, giving them this extra intelligence and the hominid ability to *reproduce* on their own.[185] So, "after the completion of the seven years, which he had completed there, seven years exactly, and in the second month, on the seventeenth day,"[186] Enki, the saurian serpent, slithered into the guarded enclosure behind his brother's back, introduced a retrovirus into the food supply, and fed it to the test subjects. As the retrovirus did its work in altering the DNA, the saurian characteristics of the original slaves began to diminish and they became more ape-like, to the point of shedding their saurian skin and taking on a similar, hairy configuration of the Neanderthals. This resulted in the need for clothing, "and He made for them coats of skin, and clothed them, and sent them forth from the Garden of Eden,"[187] to protect themselves from the elements—something unnecessary with the natural, saurian overcoat. The process took about a month, "and on the new moon of the fourth month, Adam and his wife went forth from the Garden of Eden, and they dwelt in the land of Elda, in the

---

184 A "garden" is short for a "guarded enclosure", so the "guarded enclosure of E-DIN" was later called the "Garden of Eden;" E-DIN being the Sumerian colony where the genetics laboratory (cage) was initially located.

185 The SM automatons were not able to reproduce; they were a cloned species.

186 Moses, *The Book of Jubilees*, Chapter 3, verse 17.

187 *Ibid.*, Chapter 3, verse 26.

land of their *creation*"[188] to start a Cro-Magnon slave colony, under orders of the gods to till the fields, tithe the gods, and procreate more slaves for the other gods.

Things worked well. Plenty of new slaves being created in exponential series, and the gods were happy. Until a few of these new slaves started thinking for themselves and decided to tell the gods to "get stuffed" and set out on their own, leaving the colonized areas of the gods and heading out to parts unknown, where they could not be tracked nor controlled—the "barbarians" that refused to be told what to do, and preferred to live their own lives they way they saw fit.

**Chains Without Chains**

For a moment, let's jump ahead some 5773 years since Cro-Magnon departed the guarded enclosure of E-DIN, to the 21st century, and take a look at where these slaves of the gods are, now.

The ancestral line of the Annunaki gods, what we call the "New World Order" (NWO), has successfully implemented their plans for keeping their slave society under control:

- *Physical slavery*: trapped within artificial, "political boundaries" known as *countries*, to keep the slaves as manageable groups, not allowing free access across the invisible borders without permission from the masters—and making darn sure they know where everyone is, after all, they don't want a repeat of the "barbarian" incident, where enough free-thinking men got together in one place to challenge the gods.

- *Economic slavery*: trapped within an artificial, "things of value" system, known as *money*, to keep the slaves from acquiring essential items like food and shelter, without the consent of their masters—*artificial people* known as *banks* and *corporations*. Also used as a reward system for "good slave behavior."

*Chains Without Chains*

- *Mental slavery*: trapped within an artificial "set of beliefs," known as *education*, where slaves are taught how to serve their masters, how to regurgitate what you are taught, and not question any belief, even if it is totally *backwards*.

- *Soul slavery*: trapped within an artificial "system of faith", known as *religion*, to externalize and restrict the growth of consciousness in an attempt to prevent the human soul from developing into an independent spirit.

As you can see, conformity and obedience is *rewarded*; difference (anti-social behavior) and independent thought is *punished*. People are still following the dictates of the gods, except these days we call them "lawyers," "politicians," and "priests," to name but the few. But not much has changed… obey or you will be punished. It doesn't matter if there was an actual crime or not, as the laws are full of "victimless crimes" that do nothing but attempt to legislate morality. We are *still* slaves to the

---

188 *Ibid.*, Chapter 3, verse 32.

children of the gods, and curiously enough, the *same*, genetic line of Annunaki descendants.

"… we should be trained from birth that we should all do what society wants us to do rather than what we want to do for ourselves; that because they have the technology to do it, no one should now be allowed to have their own individual personality."[189]

**Not So "Original" Sin**

It all started there, back at the guarded enclosure of E-DIN, when Enki boosted man's intelligence— fully aware that intelligence, in Medieval Latin, is the *animus*— which means *spirit*. The concepts refer to the same thing, *intelligence = spirit*. Every life form has a soul,[190] as that is part of the biological life unit. But only the more advanced life, what is termed "3rd density," has a *spirit complex*. (2nd density is just a mind/body complex.)[191] The spirit complex is present at the onset of individuality and selfidentity, that very trait that manifested in the "barbarous" rebels of the early, Annunaki colonization.

Enki knew from the beginning that this genetic change would put their slave society onto a path that would eventually lead to a new, independent and intelligent species in the not-too-distant future. His father, An,[192] was not too happy with that idea, but figured they had plenty of time to deal with it. His militant brother, Enlil,[193] was furious with this "fall" of mankind, saying that Enki had polluted the genetic line of the slaves, and all man was contaminated with this "original sin" of genetic corruption.

---

189 Monast, Serge, "Project Blue Beam," 1994.

190 The *soul* (or *mind*, in the context of mind/body/spirit), represents the cosmic (antimatter) aspect of the *life unit*, as defined by Dewey B. Larson in his book, *Beyond Space and Time*.

191 Phoenix III, Daniel, "Extra-Dimensional & Extra-Terrestrial Entities," for a description of densities and dimensions.

192 An, Anu, or El was the Titan father and leader of the Annunaki (the sons of An) or Elohim (sons of El), depending on the cultural reference. An/El was the leader of the expedition; the Annunaki/Elohim were his pure-blood descendants.

193 Enlil was also known as Jehovah, Zeus, Jupiter, Odin, Wotan, and a host of other names, after displacing An for control of Earth. He is the big "god" that is in to obedience, punishment, worship and killing everyone that challenges him, with a preference for beam weapons, the "lightning bolts" he throws.

"And so it begins…"[194] the necessity to make sure the slaves *stay* slaves to the gods, and not evolve into their own, intelligent species. A plan originally spawned by the gods, implemented by the sons of the gods, and enforced by their "royal, blue blood" descendants—those wonderful folks we now refer to as the New World Order. Things worked well for centuries, until the advent of the "printing press, v2.0"— the massive publishing capability of the World-Wide Web, where all these patriot and sovereign "barbarians" were able to get together to free humanity from the despotism of the royal lines.

Failing to achieve control over the Internet hardware and with ongoing (as of yet, unsuccessful) attempts to control "surfing" of the web, they targeted the weakest link in the system, the community of social media users, and started introducing all sorts of "insiders" with promises of freedom, ascension and free junk, cell phones, MP3 players, replicators, RVs, bank handouts, and the like, in hopes that the resulting viral spread of these ideas would overwhelm any real, pertinent information. And that plan has been working quite well, to date.

But these slaves that would not be assimilated and were still demanding freedom, created a major problem for the world leaders—one that needed to be addressed, and addressed in such a way as the slave would become *willing* slaves, once again. A condition of *voluntary servitude* is a much more desirable situation, as it becomes self-regulating; those in service will do all the dirty work for the elite, as they will do almost anything to keep from losing their handouts.

---

194 Ambassador Kosh, *Babylon 5*, episode "Chrysalis."

## Project Blue Beam

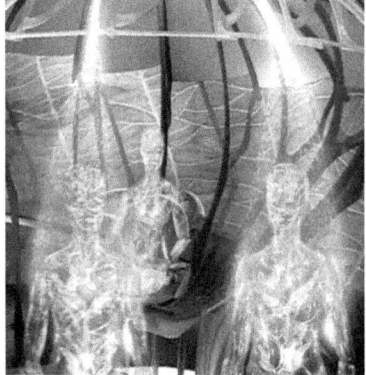

Back in 1994, Canadian journalist Serge Monast (deceased) wrote an exposé on a secret NASA project known as *Blue Beam*. These Air Force types do like the blue symbolism, as they deal with the sky. But let's take a look at some older symbolism, first, to see if we spot any other meaning.

In J. E. Cirlot's *A Dictionary of Symbols*, "Blue is the darkness made visible,"[195] while "beam" comes from the Latin *radiare*, to radiate, or *radio*. So the "blue beam" is to use some radiative process to make something appear from the darkness. With that in mind, let's examine what Monast revealed, before he "died."

*Illustration 1: The Taelon*

*Earth Final Conflict*   Monast's dissertation broke *Blue Beam* down into four sections:

### 1. Engineered Earthquakes and Hoaxed "Discoveries"

To use the archaeological version of *false flag* operations to "discover" buried, ancient tablets and artifacts that were prepared in advance and hidden so they could be found by "useful idiots"[196] after a seismic event dislodged them. These tablets, scrolls and artifacts reveal that existing religious doctrine is wrong, and "God's words" are lies.

It is well known that if you want to pull off a massive deception, always include some *truth*, and this is the case here. Yes, our religious doctrines have been greatly edited since their inception, done by those in power (politically or religiously), to stay in power. Those many people that have studied the ancient texts and done some of their own translations, realize that *all* religious doctrine stems from the same source: the Sumerian gods. Only the names have been changed to protect the guilty, and cause divisions of faith where there would otherwise be none. After all, if religious fanatics realized that they are all worshiping the *same* god, why would they ever engage in Holy Wars?

However, these "discovered" texts will *not* tell you of a common, religious ancestry, as that would *unite* people into a one-world religion that the NWO *could not claim ownership of*, and not be able to

position themselves as "intercessors" between you and your god/gods—a position they like to be in, as you, accepting them as the voice of god, will do what they say, *without question*. That's "faith." They want *willing* slaves.

---

195 Cirlot, J. E., *A Dictionary of Symbols*, Dorset Press, 1971, p. 54.

196 Wiktionary: "One who is seen to unwittingly support a malignant cause through their 'naive' attempts to be a force for good."

So what will these earth-shaking discoveries reveal? To determine that, one must examine the *desired outcome* of the "discovery."

The current movement against the NWO (or "Cabal") is being driven primarily by *Christians*, and the Christian doctrine is currently the largest in the world, at 33.39%.[197] If these discoveries were to prove the Christian doctrine is *correct*, that would solidify the anti-NWO movement, and the New World Order will end up in the FEMA camps they created to put "we barbarians" in.

However, if they were to discover that Christianity is *false*, very few Christians would believe it, and again, may result in more anti-NWO patriotism. So one of the obvious goals is to disrupt the Christian doctrine to break up their leadership of the freedom movements. To do that, they are going to need an army—and not one from the United Nations, but from a "competing doctrine."

Number 2 on the list of major religious groups, holding 22.74% of the world belief, is *Islamic Muslims*. What a surprise, looking at the current, American Presidency.[198] Put on your thinking caps and do a "what if?" some ancient discovery proves that Islam is the *correct* doctrine, and Christianity— including the Patriot movement—is nothing but a bunch of heathens. Holy Wars, Batman! With 56.13% of the world population trying to kill each other off, that should address that "depopulation agenda" they also desire. Two birds with one stone—a faked, stone artifact, that is.

## 2. The Big Space Show in the Sky

"War on Earth is fast becoming obsolete. It will be replaced by war between planets. It would do good for every nation on Earth to *unite together* in order to form a common front against possible attack by people from other planets."[199]

Monast's second phase was the "return" of the ancient *Holy Men*, "as is" for the less technically inclined, or as "ETs" for those that have watched *Star Trek* and know about holodecks.[200] This would be done using the actual blue beam hardware to

---

197 Wikipedia on "Major religious groups," from the 2010 edition of *The World Factbook*.

198 Barack Hussein Obama, aka Barry Soetoro.

199 A quote allegedly made by General Douglas McArthur, as stated on the British Sci-Fi series, *UFO*, episode "The Dalotek Affair." Of course, if he did make that statement publicly, it would be a bit of an admission on what they had planned, and as we all know, governments tend to live in the State of Denial.

200 A holodeck is a computer-generated, artificial environment that is virtually indistinguishable from the "real thing." 29 O'Loughlin, James, "Method and device for implementing the radio frequency hearing effect," United States Patent #6,470,214, October 22, 2002.

produce three-dimensional holograms in the sky, coupled with technology that allows the transmission of voice (in any language) directly into a person's skull. This technology has been around for many years, and was even demonstrated by Jesse Ventura on *Conspiracy Theory*. And as we all know, technology you see demonstrated on television is the *obsolete* stuff; they are well past sending "channeled" biocommunications from Ming the Merciless into people's brains (also known as "synthetic telepathy"[29]) and can now transmit the full spectrum of *feelings*, *thoughts* and *stimuli*, to generate the responses they desire in an untrained mind.

Imagine, if you will, a gun that shot "anger," instead of bullets. Point it at someone, pull the trigger, and that person will experience the emotion of anger. How will the brain respond? Since you are feeling anger, obviously there *must* be something for you to be angry about, and it isn't the mysterious stranger pulling the trigger that you have no awareness of. Your mind *will search for something*—anything— that it can use for an "excuse" to justify that feeling of anger, whether it be some perceived injustice from your boss at work, or your spouse spending too much money on shoes. The feeling is there, so the brain *will find something* to explain it. Then you *will act* upon that anger, causing harm or damage. A bullet may only injure a single person, but someone influenced by rage could cause a riot—a significantly more effective bioweapon. And you remain completely ignorant of the actual source. The ability to "shoot emotions" is a powerful technology and one that can be used to make a staged experience very convincing, because you "feel" it is "true," rather than determine it through rational processes (even though feelings are, technically, a rational valuing system.[201])

As Monast states, "Enough truth will be foisted upon an unsuspecting world to hook them into the lie. Even the most learned will be deceived."[5] Of course, if you've been reading my other Papers, you know that the "most learned" usually have everything *backwards*, so these new "truths" will just be an extension of more "backwards" information, to keep the "learned experts" on television, convincing everyone of their truth.

At the culmination of the "big show in the sky," the projections of Jesus, Mohammad, Buddha, Krishna, Matraia, Maitreya, Drake and others, will all merge into one—after the "experts"[202] provide the correct explanations of the mysteries

---

201 Dr. Carl G. Jung defines both *thinking* and *feeling* as rational, valuing systems attached to the "irrational" valuing systems of sensation and intuition, respectively. Within the context of the *Reciprocal System*, *sensation* is associated with the 3D, spatial senses, interpreted by thinking, and *intuition* is associated with the 3D, temporal senses (psychic ability), interpreted by feelings.

202 *Expert*, definition: A person who knows more and more, about less and less, until they know everything about nothing. 32 See the research of Dr. Judy Wood, http://www.drjudywood.com/ regarding the technology used to

and revelations disclosed, so they are interpreted correctly.

According to Monast, "This one god will, in fact, be the 'Antichrist,' who will explain that the various scriptures have been misunderstood and misinterpreted, and that the religions of old are responsible for turning brother against brother, and nation against nation, therefore old religions must be abolished to make way for the new age, new world religion, representing the one god Antichrist they see before them." As mentioned, a good deception always has a bit of truth, and the truth here is that the scriptures *have* been misinterpreted—deliberately—to keep humanity divided and conquered through the artificial boundaries of faith. The deception is that the problem is *religion*, not the *religious leaders*. Most religions tell you to "love thy neighbor;" it's the religious *leaders* that tell you to murder him.

One of the big unknowns is exactly how many people will be suckered in to this light show. There has been a lot of Science Fiction around since the inception of *Project Blue Beam*, so there may be quite a few people that question the special effects, flooding YouTube with videos of "pixel errors" on the projected face of God. So there is a backup plan, if not enough people buy the "company God" line… a taste of Armageddon: those "messiahs" get revealed as demonic ETs whom let loose the dogs of war upon a *suspecting* people, via the use of our own "Star Wars" program, the Strategic Defense Initiative (SDI). This was successfully tested on the World Trade Center.[32]

But fear not, for out of the ashes will arise a super-secret government agency, as it does in virtually *every* Sci-Fi movie, paid for by trillions of your tax dollars that has super-secret technology that can defeat the rampaging, holographic aliens, win the war, and stand ready as "heroes" so the masses can bow and scrape before them, again as willing slaves. Of course, you only need holographic phasers to defeat holographic aliens, so the technology "revealed" may also be more of a light show, than a reality.

This phase of the project has some options:

1.    People accept the return of the ancient messiahs (not likely in industrial societies), go through their raptures, and become willing slaves to the "god appointed representatives" in the NWO.

2.    The messiahs turn out to be "friendly ETs," here to harvest mankind with a mass ascension to higher states of being… they are here "To Serve Man"[203] and hand out all sorts of free toys, like Star Trek replicators and free energy devices.

3.    The messiahs are cast as demonic, evil ETs, that blast the large cities into ruins and send the people scrambling to the New World Order

---

destroy the World Trade Center and nearby buildings.

203 "Wait! It's a cookbook!" *The Twilight Zone*, "To Serve Man," 1962, based on Damon Knight's short story.

for protection.

They have done a good job setting up their win-win-win scenario.

### 3. Psycho-Terrorism: Artificial Thought and Communication

"Clearly, psychotronic weapons already exist; only their capabilities are in doubt. That is not to say that problems do not exist with the weapons and the concepts. At the present time, unpredictable systems failure and difficulty in controlling testing are major weaknesses."[204]

Psychotronic weapons are actually *biological* weapons, as they are used to target biological systems, in particular, the brain and its "software," the mind. The range of concepts is wide; for the *Blue Beam* applications, the form of "psycho-terrorism" they need is to invoke *emotional states* that are associated with the archetypal images being projected. Most people will trust their feelings over rational thought, so if God is up there trying to make logical *Illustration 2: The Visitors Provide Bliss for Humanity* arguments for His existence and return, most

*From the television series "V," 2009.* people will not believe it. But, if they are filled with joy and bliss, or depending on the scenario,

fear and terror, they will *react* before they *think* and consciously *act*. *Reaction* is *predictable* and *programmable*. *Action*, a *free will* choice is not. As to the kind of programmed reaction, Monast continues:

"Naturally, this superbly staged falsification will result in dissolved social and religious disorder on a grand scale, each nation blaming the other for the deception, setting loose millions of programmed religious fanatics through demonic possession on a scale never witnessed before."[5]

Of course, the sudden, unexpected arrival of godlike aliens from a vastly superior world would be quite a shock to human society, so society had to be *prepared* for such an arrival, even if it was all faked with fancy, human technology. In order to do so, new information would have to be presented to "civilized" man at an early enough age for them to consider it, and remain in social media sufficiently long so they teach their young to accept it. This normally requires two *generations*,[205] where

---

204 Alexander, Lieutenant Colonel John B., "The New Mental Battlefield: Beam Me Up, Spock," *Military Review*, December, 1980, page 53.

205 The period of time for a newborn to grow up and create another newborn.

the first generation is introduced to the concept, gets to live with it for long enough that it becomes safe and mundane, then pass it along to their progeny, whom accept it as "matter of fact." In industrialized countries, the human generation is about 25 years; the time it takes for a newborn to grow up, get fully programmed into society, reproduce, and educate their young. So a proper preparation of this kind of event needs planning and about *50 years* to execute.

Currently, society is very accepting of "aliens" and the ET concept, because it has been around since the hippie movement in the mid-1960s. Guess what… that's 50 years ago. Society has been fully "prepped" and is ready to be fully "conned."

The quotation made by Lieutenant Colonel Alexander starting this section was made in *1980*— psychotronic weapons were well under way by then, including "synthetic telepathy," the use of transmitters to put voices directly into the heads of unsuspecting people. Curiously enough, the early 1980s was also the start of the "channeling" craze—no longer the typical mediumship of the earlier years communicating with the spirits of the deceased, but now with a new twist: aliens from outer space, other planets and other dimensions.[36]

Monast's inferences are that all this channeled information, purportedly from extraterrestrials, is nothing more than *propaganda* in an attempt to control the spiritual development of man by keeping the "god" concept externalized—simply shifted over from the old gods, to the new ETs. But still the same group pulling the strings. The development of synthetic telepathy could not have happened at a more opportune time, as it coincided with the "recreational drug" craze that breaks down many of the mental inhibitions to radical ideas. Or, perhaps, the drug craze was *introduced* at the same time to assist in the acceptance of this new form of "C3."[37] Regardless of which came first, the result was the same: *success*. They were able to introduce all sorts of "new age" concepts into a willing population, for an agenda that was planned for execution some half a century later. If anything, the New World Order are *patient* folks.

### 4. Universal Supernatural Manifestations via Electronics

"Seeing is believing," right? If you are making plans to control the spirit of an evolving species, that necessitates being able to create *convincing imagery*—not just in the sky, but in the minds of those you wish to control. It is not a new theme; it has been used on the BBC series, *Doctor Who*, a number of times, from the "ghost shift" from the Season 28 episode, "Army of Ghosts,"[38] all the way back to the 1968 episode, "The Invasion,"[39] where the Cybermen added a

*Illustration 3: The Ghost Shift*      "micro monolithic circuit" to the then popular

transistor radio that

---

With the current state of "delayed

*Doctor Who, "Army of Ghosts"* allowed them to take control of human minds, via the "cyber control signal" beamed from space.

---

adolescence" in civilized societies (do not start a family until out of college), the generation is approximately 25 years.

36 Actual extraterrestrial contacts with "regular" people were jammed in the late 1950s, as the contacts were usually radio based (typically Morse code)—*direct* contact, *not* telepathic or channeled.

37 *C3* is a military term for "Command, Control & Communication" of human activity.

38 *Doctor Who* (#10, David Tennant), "Army of Ghosts," written by Russell T. Davies. Aired 1-Jul-2006. The ghost shift was a technology used to transmit images from a parallel universe to the present one, that had the appearance of ghosts.

39 *Doctor Who* (#2, Patrick Troughton), "The Invasion," written by Derrick Sherwin from a story by Kit Pedler. Aired

2-Nov-1968 through 21-Dec-1968. The Cybermen used hidden electronics to hypnotically control the world population.

That old *Doctor Who* episode actually had the right idea. To introduce subconscious signals into any kind of media stream, you do not embed it in the *media* where it could be discovered or filtered out, but *into the equipment* used to express that media directly into the brain: MP3 players, cell phones, computers, televisions, radios and the like. These days, they are all "micro monolithic circuits," whose functions are buried deep at the atomic level of integrated circuitry, so you could put anything in there without the public ever becoming aware of its function. When it comes time to activate, all one would have to do is to broadcast that "cyber control signal" across the global, cellphone network, and out the 6 *billion* cellphones where it will not only effect the owner of the phone, but most likely many others in the immediate vicinity. Back in 1968, a *two transistor* radio was "hot stuff." A typical, multi-core CPU chip these days has over *two billion* transistors. That's a lot of logic, and as it is said, "the best place to hide a tree is in a forest."

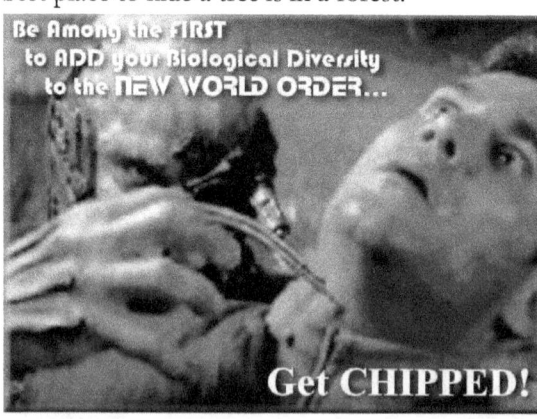

Knowing that the technology is not only available, but *implemented* globally, allows us to examine Monast's final section in detail, revealing that the technology is currently available to "rewrite" the software of the human brain via electronic devices in close proximity. Before long, *Blue Tooth*, or should I say, *Blue Beam Teeth*, will be taking another big bite out of freedom, controlling the *electronics* of your mind. And while you are at it, don't forget to have your kids "chipped," so "you" know where they are, at all times. After all, it's for *your* benefit and has *nothing* to do with global domination and willing slavery.

Monast defines what he believes will be the final stages of the process of taking control, not only of the body and mind, but of the newly-evolving spirit:

> 1. "It will begin with some kind of worldwide economic disaster. Not a complete crash, but enough to allow them to introduce some kind of inbetween currency before they introduce their electronic cash to replace all paper or plastic money."[5]

It does not take much "surfing the net" to discover that this is already well under way, with projects like

OPPT (One People's Public Trust), Neil Keenan and his "global funds," revaluation (RV) of Iraqi Dinars, free "replicators" for the good slaves to fill their lives with material possessions, "reset" of the banking system to remove all the debt you've run up with uncontrolled spending… the list just goes on and on, and the one thing they all have in common: they are all *handouts:* you get *something for nothing.* Nature uses the "handout" concept frequently—the only difference is that Nature calls it: *bait.*

> 2. "The second is to make the Christians believe that the Rapture is going to occur with the supposed divine intervention of an alien (off-world) civilization coming to rescue earthlings from a savage and merciless demon. Its goal will be to dispose of all significant opposition to the implementation of the New World Order in one major stroke, actually within hours of the beginning of the sky show!"[5]

The *same pattern*, except instead of handouts, we have the "savior," another old concept where one entity is burdened with the job of saving the *helpless* people, via *intervention*, the "mass arrests" of a corrupt, political or commercial system, or *rescue* via "harvest" or "mass ascension" to a higher state of being where you escape the system of oppression. Again, it requires you to do nothing except kick back, pop another brew, and "enjoy the fireworks." Convincing people that they are helpless is a good way to reduce opposition.

> 3. "The third orientation in the fourth step is a mixture of electronic and supernatural forces. The waves used at that time will allow 'supernatural forces'[206] to travel through optical fibers, coaxial cables (TV) electrical and phone lines in order to penetrate to everyone at once through major appliances. Embedded chips will already be in place. The goal of this deals with global Satanic ghosts projected all around the world in order to push all populations to the edge of hysteria and madness, to drown them into a

---

206 Communication through 3D time, the cosmic aspect of the life unit where feelings and intuition dominate. 41 *The Thomas Gray Archive*: "Ode on a Distant Prospect of Eton College" (Lines 99.3-100.5)

wave of suicide, murder and permanent psychological disorders. After the *Night of the Thousand Stars*, worldwide populations will be ready for the new messiah to re-establish order and peace at any cost, even at the cost of abdication of freedom."[5]

This is the natural consequence of a fear-based society; to push the level of fear and misunderstanding so high, a person cannot operate within their own world view and then takes their own life, or goes on an uncontrolled rampage—a form of "assisted suicide;" the ultimate *Manchurian Candidate* that did not even need to be put through a mind control program—they just make use of the hopes and fears already in place, through the careful nudging of mass media. If a *willing slave* is "set free," they will do *anything* to find a new master to bring meaning and purpose back to their life.

**The Mind Has No Firewall**

In the old days, a "firewall" was just a noncombustible wall designed to protect a structure from fires. That wall could either confine the fire to a specific building or protect a building from an outside threat. Computer "firewalls" do the latter; it acts as a barrier surrounding your computer to protect it from external damage in the guise of computer viruses, unsolicited commercial advertising (SPAM) and the like. Your brain and mind are much like a computer and its software, where by default, the "mental firewall" is *switched off*, allowing itself to be constantly infected from the outside, from the direct programming of "education," to the subtle mechanisms of subliminal advertising, synthetic hypnosis, subconscious suggestion and a myriad of other processes.

However, these processes are only effective when you are *unaware* of them. The purpose of this Paper, as well as my other Papers on a variety of topics, is *to make you aware* of what is potentially going on around you, so *you have the option* to switch that firewall *on*. And should you switch it on, like any piece of new software, it will take some "tuning" to get it working properly, filtering out the "cyber control signals" attempting to influence your behavior. But remember, that in our society, "where ignorance is bliss, 'tis folly to be wise."[41]

Will taking up arms against this technological oppressor accomplish anything? In my opinion, *no*. "A leader does not need guns or knives to give him authority. His weapon is *intelligence*."[207] Turn on your mental firewall. Stop being subconsciously influenced. *Become intelligent*. Know what you do, and why you do it. Then you will see that there is another way, a "natural consequence" of biological evolution that has been slipping between the cracks, where those in power hope you will never look. As Roger Damon Price described it in a British television series back in 1973, the future of mankind are *The Tomorrow People*.[208]

---

207 Sister George (Stella Stevens), *Where Angels Go, Trouble Follows!*, Columbia Pictures, 1969.

208 Price, Roger Damon, *The Tomorrow People*, Thames Television for ITV Network, 1937-1979.

**Resistance is Futile; Evolution is Effective**

In Price's series, *The Tomorrow People*, young people developed ESP, "extra sensory perception" or "psychic ability," what we call nowadays, *psionics*. These *homo novus*[209] shared a specific world view centered around personal evolution to become peaceful explorers of the universe. As an impressionable teenager at the time, it was one of those public television shows that my friends and I would rush home to watch, as it gave us hope that there *just might be a chance* that we could grow up in a world that worked together for the betterment of mankind—even if *children* had to lead the way.

Within the framework of the *Reciprocal System* of theory, psionics are a *natural consequence* of the continued evolution of life units, creating a new, third level of existence for mankind (the first level being the *inanimate*, and the second being *biological* organisms). Because of this *theory of everything* basis, man, left to his own mechanisms, will *eventually outgrow* the need for competition, rivalry and fighting, and evolve into an *ethical being*, with the tools, knowledge and compassion necessary to work in rapport with all life. Some are already standing on that threshold, however there are sinister forces at work, as described earlier, that are *consciously* trying to prevent this evolution of mankind. As Larson describes, ethical man is an evolving subset of Cro-Magnon man—not many, but some:

"It is therefore evident that we cannot equate man with the Level 3 structure in the same manner that we were able to equate life with the Level 2 structure. Rather, we will have to identify the Level 3 structure with an idealized kind of human: an *ethical man*, let us say, giving the term "ethical" a very broad meaning. The boundary line between Level 2 and Level 3, then, is not between *animal* and *man*, but between *man* and *ethical man*. However, much of the human race is partly across the boundary; that is, each of these many individuals is at some times, and to some degree, under the domination of the Sector 3 control unit rather than the Sector 2 life unit."[210]

Larson continues to analyze his "Level 3" and the influences of the Sector 3 *control unit* that creates an ethical consciousness, determining that this is where mankind exhibits his finest features: rapport, compassion, moderation, self-sacrifice and the conscious use of psionic skills, since both halves of the universe, the physical/material and the metaphysical/cosmic, are available to consciousness. When someone, of any age, comes under the influence of these Sector 3 control units, their motivations switch from the biologic *rivalry of competition*, to the ethical *rapport of cooperation*: they become a Tomorrow Person. And it is a "no turning

---

209 Originally *homo superior* in the series, later changed to *homo novus* (new man) to better describe the Tomorrow People as the next stage in human evolution, rather than a Hitler-style superior race.

210 Larson, Dewey B., *Beyond Space and Time*, North Pacific Publishers, Portland, OR, 1995, page 82.

back" situation—once you access that level as part of your being, you begin to *understand* things for what they are—not just *see them* as presented to you, which is why the New World Order folks must prevent this evolution from happening. It cannot be turned back once it does, so as long as you are following their artificial paths to ascension, trapped within their artificial boundaries, resistance *is* futile. Cross the lines that aren't actually there, and you'll find evolution *is* effective.

### The Boundaries of Your Mind

The boundaries discussed in the "Chains Without Chains" section are psychological constructs, programmed into you at birth. They are **not** *natural* boundaries. Forget about them for a while; forget your faith, your patriotism, your corporate loyalty. When you bring those walls down, you can get a look at what it is that has been *blocked* from your consciousness by *The Powers That Be*. Knowledge is power, and the most powerful knowledge are the secrets behind those artificial walls in your mind.

In case you hadn't noticed, I am a proponent of Dewey Larson's *Reciprocal System of theory*, which is technically a "TOE," a *theory of everything*. And it is a simple one, using natural consequences from a ratio of space to time, called motion (though I am the first to admit that Larson does not do a good job explaining his ideas). It is not a popular theory because it is a consequence of *nature*, not of *mathematics*. Conventional science demands a TOE that *proves that they were right*—not one that says they screwed up from the start and spent the last 300 years making a total disaster of everything.

When trying to learn Larson's concepts, one spends more time *unlearning* what they've been taught, as they were taught everything *backwards*. And that's the key to dealing with the evolution of the human spirit—to throw a monkey wrench into the plans of the New World Order—just jump their walls and you'll find that mankind, a bizarre, interplanetary slave hybrid, has more potential than any other species in this part of the galactic neighborhood. And believe me, our neighbors are well aware of that fact.

Since Project Blue Beam is primarily focused on extraterrestrials and aliens, let's *unlearn* a few things about "who's who" and "what's that" in the neighborhood.

### Unlearning: Breaking Down Artificial Barriers

In *The Universe of Motion*,[211] Larson outlines a process to construct an *entire universe* that only requires two things: *gravity* and the *cosmic microwave background radiation* (CMBR). The process is simple; gravity aggregates radiation into particles, particles into atoms, atoms into aggregates, aggregates into stars, stars into galaxies and so forth. Gravity is a *natural consequence* of matter, so all you really need is a *constant influx of matter* from somewhere. Larson identifies the origin of the CMBR as *not* a "leftover from the Big Bang," which one would think would have run out of juice by now, but from the *cosmic sector*,[212] the realm of 3D time known in

---

211 Larson, Dewey B., *The Universe of Motion*, North Pacific Publishers, Oregon, USA, 1984.

212 The *cosmic sector* was named as such, because it was identified as the

conventional science (and science fiction) as the *universe of antimatter*.

Stellar and galactic combustion processes are based on "age limit" fission in the *Reciprocal System*, which are constantly pushing matter to faster-than-light speeds, as exhibited by RF and X-ray emissions[213] of both. Once you move faster than light, you are moving in 3D time, not 3D space, so our half of the Universe is sending over it's own "*material* microwave background radiation" to the cosmic sector, where it built a cosmic universe in 3D time, which is always sending us back its "antiradiation" as the CMBR. That's the *reciprocal relation* at work, and why the *Reciprocal System* is named "Reciprocal." It is always a constant exchange.

So, the first lesson to be unlearned is that the CMBR is a *constant influx of matter* from 3D time. As such, it was *not created* by a Big Bang, and since *nothing else* is needed to create a Universe, there was *no* Big Bang needed, either!

So when ETs or channelers tell you all about them being the "best bang since the Big One,"[214] they are lying. They are stooges for the NWO, trained by those conventional scientists that believe in the Big Bang theory, or just "useful idiots."

*Unlearning: Breaking Down Artificial Barriers*

Stars, as described by Larson and referenced in my other Papers, are constructed from this dust forming from the CMBR. That means they start out cold, warm up, get hot, hotter and extremely hot, until they blow apart in a supernova. The existing stellar evolutionary sequence is backwards, and since that is the basis of galactic evolution, that is backwards, also.[215] And remember—all you need to create everything in this universe, is just *gravity* and the *CMBR*. Knowing the correct stellar evolutionary sequence allows us to learn of how stars evolve and produce planets, how planets produce life—and hopefully, ethical life.

A substantial part of *The Universe of Motion* is dedicated to stellar evolution. Summarizing, we find that stars grow in *generations*:

1.    First generation stars are newly formed from dust and debris and have no planets. Any planetary matter in the vicinity would be sucked up into the new star, adding to the stellar mass. These stars are characterized by a single star, with no companion, and "clean" space around them.

2.    Second generation stars form after the first generation star undergoes a supernova by getting too hot (Larson's *thermal limit*). The result is a *binary* star system: a red giant that is composed of the matter in

---

source of cosmic rays and radiation by Larson.

213 Larson, Dewey B., "Astronomical X-Ray Sources," *Reciprocity* V, № 1, page 3.

214 Gallumbits, Eccentrica, *The Hitchhikers Guide to the Galaxy*, opinion of Zaphod Beeblebrox.

215 A consequence of the reverse galactic evolution, the Andromeda galaxy is much *younger* than the Milky Way.

3D space and a white dwarf, where the old solar core imploded in space and is expanding in 3D time, giving it its compressed form with an inverse density gradient.[216] Over time, these binaries move to the main sequence and repeat the supernova cycle.

There is a *possibility* of a solar system at this time, if the stellar core is spatially fragmented during the supernova. However, the white dwarf fragments would be small, cool off quickly, and most likely get sucked back into the sun while still in its giant phase, so they would be unlikely to produce planets having intelligent life.

3.      Third (and greater) generation stars are either triple star systems, or single star systems with a collection of stable planets. Third generation stars, like our sun, Sol, are the generation of stars most likely to contain planets with intelligent life. They are identified as single red to orange stars with a large amount of surrounding debris (young with forming planets), or yellow to white with an asteroid belt and planets (middle aged with established planets).

### ET Phone Home: Long Distance!

We can now use a process of elimination to find where *real* ETs are likely to reside. A single sun without a debris field, like the stars of the Hyades, won't have any planets as they are 1st generation stars. Binaries, 2nd generation, are also out, as the stuff planets are made of is still locked up in that white dwarf companion. Triple star systems, like Alpha Centauri, are also out for the same reason—the planetary cores are still in the white dwarf component.

So let's take a look at what astronomers have recently discovered and graph them in the proper direction of stellar evolution, to find out where our ET neighbors are, within a 45 light-year radius from Earth. See Chart 1.

Examining the chart, it becomes apparent that many of the popular, ET-channeled stellar systems *don't seem to have any planets around them*, being first or second generation stars. They were picked because the names are well known by many cultures, and have been for centuries.

---

216 Larson, Dewey B., "The Density Gradient in White Dwarf Stars," *Reciprocity* XI, № 2 (Summer, 1981).

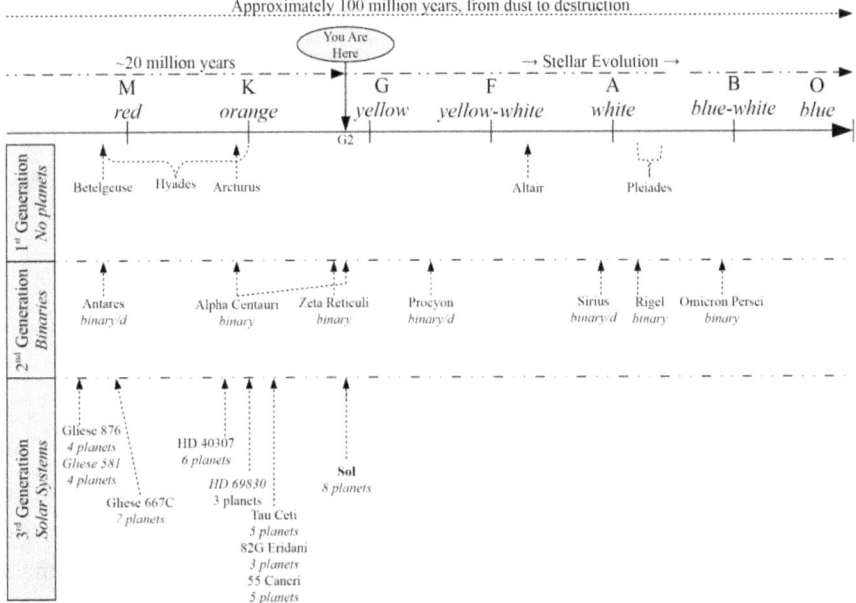

*Chart 1: Solar Systems within 45 Light-years*

Something becomes apparent: Sol, our solar system, is the *most evolved* solar system within this 45 light year range! We're the most advanced kids in the neighborhood. Tau Ceti, a G-class, yellow star much like our own, is a neighbor with planets— but those planets are *younger* than Earth, perhaps still back in the Jurassic period with dinosaurs running around. That also means a high magnetic ionization level, so the planetary environments would be *radioactive*—a similar problem the SMs faced here on Earth, during their colonization.

And yes, our sun is only about 20 *million* years old, according to the calculations of Prof. KVK Nehru of India, basing his computation solely on gravity and the available amount of matter in the area, including the CMBR.[217] The estimated life span of a star generation, in the *Reciprocal System*, is approximately 100 million years, from dust to supernova.[218]

So where *are* the ETs hiding? The stars most likely to have intelligent life would be 3rd generation, old enough to get by the worst of the early radiation levels, but not too old as the sun would become too hot with radiation into the ultraviolet and X-ray bands that would become destructive to life. What we are looking for are 3rd generation, G and F class stars:

*ET Phone Home: Long Distance!*

---

217 KVK Nehru, "The Large-Scale Structure of the Physical Universe, Part II: Mathematical Aspects of the Cosmic Bubbles," *Reciprocity* XX, № 2, Equation 21.

218 Phoenix III, Daniel, *Geochronology, Part 1 of the Hidden Origin of Homo Sapiens*, discusses the problems with geologic dating, and how existing dates are substantially too large.

| Star | Planets | Stellar Class | Distance (ly) | Relative Age |
|---|---|---|---|---|
| Kepler-11 | 6 | G4 | 1999 | *Younger* |
| Nu2 Lupi | 3 | G2 | 47 | *Same Age* |
| Kepler-9 | 3 | G2 | 2754 | |
| HD 10180 | 7 | G1 | 127 | |
| 47 Ursae Majoris | 3 | G0 | 46 | |
| HD 1461 | 4 | G0 | 76 | |
| HD 96700 | 2 | G0 | 83 | *Older* |
| HD 169830 | 2 | F9 | 119 | *(More* |
| Upsilon Andromadae | 4 | F8 | 44 | *Advanced)* |
| HD 60532 | 2 | F6 | 83 | |
| HD 8799 | 4 | F0 | 129 | |

If I were looking for intelligent ETs in the area, my choice would be HD 10180, which is a few thousand years further along in evolution than we are, with a large, 7-planet solar system that is similar to our own. With similar conditions, similar life may evolve. But, I've yet to run across hyperintelligent, pan-galactic ETs (or even mice) that come from HD 10180, a star so insignificant our astronomers never bothered to give it a "real" name.

## Conclusions

There is an active, ongoing program to suppress the spiritual growth of humanity, being implemented by those folks that consider themselves to be the direct descendants of the gods. They have positioned themselves as royalty and political leaders, priests and gurus, and the leaders of the military-industrial complex. Like the Kings of old, they consider their "genetic purity" of ancestry to be the dictate for their rulership of the planet and the only reason the remainder of mankind is allowed to exist is to be their slaves. Their *modus operandi* is rivalry and competition, using the tools of technology.

You have been trained to believe that the "man in the mirror" is who you *really are*, having about as much substance as your reflection. The powers that be are experts at manipulating reflections and have used it to their advantage to make sure you never realize that there is something real and tangible *casting* that reflection.

Through control of education, they have you believing everything is backwards so cannot see the consequences of Nature. Even if you try to take a step forward, they have all the spiritual avenues well guarded with the "ascended master" of your choice. So, *buyer beware.*

Correcting the stellar evolution shows that our solar system is the most evolved in the immediate area of nearly 100,000 cubic light-years, and there are not a whole lot of similar solar systems nearby. But, astronomical research has shown that there are hundreds of "up and coming" solar systems circulating about red and orange stars that *will* become viable in the near future for explorers and colonists. The natural consequences also indicate that the majority of suns that channeled sources claim to be from, *do not have planets*, hence the logical conclusion is that they are *part* of this spiritual misdirection. Most of these sources also claim "mass ascension," where you can receive all the knowledge in the Universe, just by sitting on your butt sucking a brew, watching people beating up each other on sports shows. I ask, "could anyone of reasonable intelligence actually believe this?"

### Epilog: The Cro-Magnon Matrix

Just like the *progression of the natural reference system* described by Larson that causes all physical systems to grow and interact, Nature also impresses its goal to evolve consciousness on all the life that comprises it. There is substantial information indicating that man is a genetic hybrid between native life and an extraterrestrial species, but man is still *part* of the life matrix on Earth and *subject* to those same rules that desire the evolution of consciousness. That gives us an evolutionary option that cannot be taken away—only *disguised*.

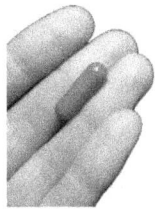 These "blue bloods" have a *blue pill* to offer, and if taken, you remain ignorant of your own potential and live in their world of artificial boundaries with backwards education, reaping the material rewards of being a good slave. Not having achieved a state of ethical consciousness, the jump to "Level 3" that Larson describes, you remain in the loop, and as any engineer can tell you—entropy increases, until there is nothing left to make the loop again.

Blue pill vendors always have a common agenda: that of *rivalry*, even if they are stooges believing they are doing the right thing. Things to look for: competition, fighting, wars, starfleets of warships, hierarchal orders, chains of command (be it political, social or military), superior beings for you to worship, "us" versus "them," free handouts, promises to get something for nothing or ascension without effort. Anything based in fear, control, power or ego gratification.

Red-blooded Terranean genetics have a *red pill* to offer, part of our inheritance from the apes and Neanderthals that make up a significant part of our DNA. If taken, you become aware of your own potential and can work to transform yourself and the world around you back to a natural course of evolutionary development. *The Tomorrow People*, welcomed as peaceful explorers in an ever-evolving Universe.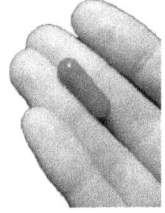

Red pill vendors also have a common agenda: that of *rapport*, of working together to accomplish common goals, where each donates their unique approach to life, without demands of compensation. Things to look for: cooperation, harmony, compassion, exploration (of any subject), psionic ability used to assist healing and evolution, requests that you do the work *yourself*, and share what you discover on your path; hard work, dedication and ascension with a LOT of personal effort![219] Using these basic agendas, it becomes fairly simple to sort through the myriad of information available, to find what is real, leading to the evolution of consciousness, and what is just a showpiece, leading back to the start after a long delay and detour.

---

219 Obvious by its absence, is the concept of *love* in this list. The emotional responses of love have been so abused and distorted by The Powers That Be that they have become unrecognizable by most of mankind. I find no benefit in using a word that no two people will define the same way.

The first step to solving any problem, like the spiritual evolution of mankind, is to identify the problem. Hopefully, this research will assist you in doing just that. Now that you have an understanding of the problem, mankind can start cooperating to find a solution to get us back on track, without having to resort to the same fear tactics that those in power are already using.

Just remember: the red pill is hard to swallow at first, then as you wake up from the illusion, you find the freedom it provides is rather sweet.

# HOMO SAPIENS ETHICUS
# LIFE, DEATH, REINCARNATION AND ASCENSION

## Introduction

In order to understand how the created, Cro-Magnon man differs from other evolved life on our world, an investigation into the structure of life, itself, is a necessary prerequisite. Once the *natural norm* is defined, deviations from that norm can be investigated and consequences determined.

This chapter will cover two basic concepts, as energetic consequences: *life* and *death*, along with the evolution of these processes: *reincarnation* and *ascension*. Life, as a natural consequence of the postulates of the *Reciprocal System*, is covered in detail by Dewey Larson in his book, *Beyond Space and Time*.[220] In hopes of defining the basic pattern of what causes *natural* death and what happens during, and after, the death of the physical body, the concepts of death, reincarnation and ascension are extrapolated from the core concepts of cultural mythology and theology, correlated to corresponding concepts in a framework of a universe of motion. This information can then be extrapolated to see where mankind, as a species, is heading.

The pretext of this paper, based on the concepts proposed in *Geochronology*[221] and *New World Religion*,[222] is that Cro-Magnon man, from which modern man is a direct descendant, is a *hybrid* of the evolving life on the planet *plus* an "extra-terrestrial" or "divine" influence that was introduced by a species collectively referred to as the "SMs" that colonized the planet in ancient times, creating the Mu and Atlantean epochs.[223] The progenitors of this hybrid species of man are commonly referred to as the Biblical Adam and Eve, so this hybrid approach is a mix between Darwinian views and theological ones—*both* are correct. Man was *created* and is now *evolving* on his own, as a distinct species.

Terminology is going to be difficult as many of the words have subjective

---

[220] Larson, Dewey B., *Beyond Space and Time*, Tucek & Tucek, published posthumously in 1996; written in 1979.

[221] Phoenix III, Daniel, "Geochronology: Hiding History in the Past; Part 1 of the Anthropology Series on the Hidden Origin of Homo Sapiens," 2013.

[222] Phoenix III, Daniel, "New World Religion: Enslaving the Human Spirit with a Blue Beam; Part 2 of the Hidden Origin of Homo Sapiens," 2013.

[223] This concept is referred to as *Intervention Theory*.

meanings to the reader. To help alleviate this difficulty, the terms will be defined prior to being used, so context can be maintained. There will be those that disagree with the choice of specific words, but these are what we've been using in the scientific underground for 30 years and have held up well. It is the *concept* being related that is important, not the word used to express it.

There will also be those who vehemently oppose Larson's concept of *consciousness* being a *consequence of life*, rather than *matter* being a *consequence of consciousness*. I ask that you "suspend disbelief" for the moment, as it *is* a reciprocal relation. Asking "what came first" is like asking if the inside or outside of a box came first—neither did. Once you have a box, you've got an inside *and* an outside. The same situation exists with consciousness; it's all a matter of perspective.

## Life

To the best of my knowledge, Dewey Larson has the only theory, the *Reciprocal System* (RS), that deduces both *life* and *ethical consciousness* as a *natural consequence* of its Postulates. What Larson proposed in his book, *Beyond Space and Time*, is that life is essentially *a stable combination of matter and antimatter*.[224] Most physicists will tell you that when matter and antimatter meet, they cancel each other out and you get an explosive result. What Larson found was that matter and antimatter can also meet "out of phase" with each other, such that they form a stable, more complex structure: the *living cell*. Still an explosion, but a *constructive* explosion of *life*, rather than a *destructive* explosion of *radiation*. Because of this mix of material and cosmic (antimatter) motions, the living cells operate primarily in the *intermediate speed* range, analogous to the processes in *stars*, but due to this stable balance, the process is at much lower temperatures. To quote Delenn from *Babylon 5*, "We are starstuff,"[225] literally.

This unique understanding of the living structure as an aggregate of a physical, spatial body and an invisible, temporal body, allows an investigation into that "other realm," the cosmic half of life that philosophers and spiritualists have identified as the *soul* or *mind* (as in the conventional, New Age concept of Body/Mind/Spirit). Whereas soul and mind have many connotations, the Medieval Latin term, *anima*[226] will be used, as commonly used by psychologists. *Anima* will be used to refer to this "soul" half of the living organism, the unobserved presence in 3D time.

---

[224] *Antimatter* in the *Reciprocal System* is called *cosmic matter* and is technically "conjugate matter," composed of atomic rotations in space, placed in a 3D temporal coordinate system, using *clock space* as a measurement of change. Larson refers to it as *inverse* matter, but it was pointed out by Prof. KVK Nehru that the inversion must actually be a *conjugate*, to preserve dimensional relationships. For example, the inverse of material force, $t/s^2$, is $s^2/t$, which is incorrect. Cosmic force has the dimensions of $s/t^2$. The aspects of space and time invert, but the dimensional relation in the numerator and denominator stay the same, making it a conjugate, not an inverse.

[225] *Babylon 5*, "A Distant Sun."

[226] "That which animates" or brings to life.

According to the *Reciprocal System*, the *physical*[227] universe consists of two *sectors* of expression, existing 90° out of phase; each the antithesis of the other; two *perspectives* of scalar motion:

1.    *Material sector (matter)*: three, coordinate dimensions of *space* with one dimension of *clock time* (duration). This is the observable, measurable sector of our everyday experience, known to the 19th and early 20th century researchers as "ponderable matter."

2.    *Cosmic sector (antimatter)*: three, coordinate dimensions of *time*, with one dimension of *clock space* (distance). This is the origin of *etheric* phenomena ("imponderable matter") and is not directly observable nor measurable *from space*. We can indirectly measure how *time changes space*, by observing how material structures change when influenced by temporal ones.

However, when confronted with certain, observed phenomena such as ethical behavior and psychic ability, Larson found that the two sectors of his physical universe did not supply sufficient relationships to account for these features and he added a 3rd, *nonphysical* sector to hold all the evidence and observations that were yet to be sorted out and placed into the theoretical framework of the *Reciprocal System*. He called this third sector the *Ethical sector*, and the structures created within it, *control units*.[228] Unfortunately, Larson died of old age before completing this study.

*Life*

Larson's concept of "sectors" also maps to the esoteric concepts of ontological *planes of existence*:

| Sector # | Sector Name | Esoteric Plane |
|----------|-------------|----------------|
| 1 | Material | Physical |
| 2 | Cosmic | Astral |
| 3 | Ethical | Causal |

Note that Larson's investigation ended with his third, ethical sector, because of his death. This does not imply that there are not additional sectors, nor levels of existence, waiting to be unraveled as natural consequences of the *Reciprocal System*.

### The Levels of Existence

Larson defines three different *levels of existence*, based on the relationships between the three sectors:

---

[227] "Physical" includes all the relationship that are defined by 3D, scalar motion with the aspects of space and time. This includes matter (material atoms), antimatter (cosmic atoms), electric and magnetic fields, gravity, biological organisms (life, including body and mind/soul, but *not* "spirit") and their bioenergetic systems.

[228] Larson, Dewey B., *Beyond Space and Time*, *op. cit.*, p. 81.

| Level | Concept | | Sector | | | Motion | Esoteric |
|---|---|---|---|---|---|---|---|
| | | | 1-Material | 2-Cosmic | 3-Ethical | | |
| 1 | Inanimate | Matter | X | | | Particle, Atom, Molecule | Body |
| | | Antimatter | | X | | | |
| 2 | Biologic | Living | X | X | | Life Unit | Mind/Body |
| 3 | Ethical | Metaphysical | X | X | X | Control Unit | Mind/Body/Spirit |

1. *Inanimate*: Material **or** cosmic sectors; the atomic and chemical realm; matter **or** antimatter. Combinations *destructively* interfere to produce energy.

2. *Biologic*: Material **and** cosmic sectors, combinations *constructively* interfere to produce cells. The material, spatial aspect of life is the *corpus*, the body. The cosmic aspect of life is the *anima*. Both halves, together, are referred to as a *life unit*.

3. *Ethical*: Material **and** cosmic **and** ethical sectors; a structure built upon the framework of space and time, but reaches beyond it. These "ethical control units" are *cells of intelligence*,[229] that we collectively refer to as *consciousness*. The influence of these cells upon the body, mind and soul, as the *spirit*, for which we will use the Medieval Latin term, *animus*,[230] to eliminate preconceived notions concerning the concept of the spiritual.

As can be seen, Larson, through the *natural consequences* of his *Reciprocal System*, is slowly and surely beginning to deduce the existence of the esoteric, philosophical or spiritual realms—purely from the relations of space and time as motion. And he is finding that they have the same structure as ancient literature suggests, with the only difference being that as natural consequence of the theory, the attributes and properties of these levels, sectors and units can be *explicitly determined*—as well as how they interact with each other. This gives us the basis to analyze evolving life, as well as the concepts that go along with life, such as *bioenergy* (prâna, qi, ch'i, shen, kundalini and other forms), *death, reincarnation* and *ascension*. We don't have to *guess*, when we can just *deduce* where life is going, how it is going there and what we can do to assist this growth of consciousness.

**Force and Energy**

We are already familiar with the basic *forces* of Nature, namely *electrical, magnetic* and *gravitational*.

We also know about certain forms of *energy* as *waves*, of which there are two ranges in the *Reciprocal System*, divided by unit speed:[231] *Low Frequency* (LF) waves that contain low energy and do little harm, such as sound, light, heat and radio; and *High Frequency* (HF) waves, which can cause serious injuries, such as x-rays, gamma

---

[229] Not to be confused with the *moron*, a particle of stupidity that is highly contagious in bureaucratic systems.

[230] *Animus* is typically defined as the intellect or spirit, often associated with the concepts of ethics and morality.

[231] *Unit speed* is the speed of light, 1 natural unit of space per 1 natural unit of time, *unity* in the Reciprocal System, or 299,792,458 m/s in conventional physics.

rays and cosmic rays.

Force and energy, though similar, are *not* the same thing. When Larson's *natural units* of space (s) and time (t) are used to define quantities (rather than the names of dead scientists), the concepts behind force and energy, and their differences, become obvious.

Energy (t/s) remains constant, but *force* (t/s / s = t/s²) is energy that *changes* with respect to *space—not time*. If one looks at the conjugate relationship, speed that changes with respect to time, s/t / t = s/t², we find this is a familiar concept: *acceleration*. Force is just *temporal acceleration*. And because coordinate time is not directly observable, it acts *invisibly* on space.

This gives a clue as to the nature of energy and why we can only detect it by the way it *changes space*: force and energy are *temporal* structures, moving with *clock space*, not *spatial* structures moving with *clock time*. As temporal structures, they exist in 3D time (or as physicists say, are *localized* in time).

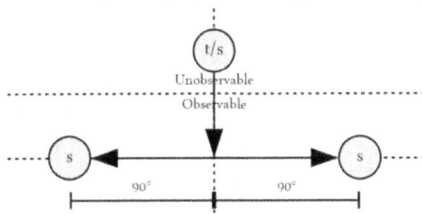

This is easily observed by the way a magnetic field moves iron filings, making the shape of "lines of force." We cannot *see* the magnetism itself, because the effect is nonlocal, but we can see *how* the magnetism changes the spatial arrangement of the iron filings.

It is the temporal component that is the "energy," t/s, which causes a change in space, Δs; the observed result of a *force*, t/s / Δs = t/s².

All *life units* contain atoms, atoms have electric and magnetic fields and are responsive to gravity, just like their inanimate counterparts. But life goes beyond the electromagnetic forces due to the linkage between the material and cosmic atoms—that stable, matter-antimatter relationship—and that relationship generates additional fields expressed by the term, *bioenergy*, the energy of biologic organisms. Keep in mind that everything that happens in our observable, spatial reference, also happens in the unobservable, temporal reference, resulting in additional "anti-" concepts, such as *antigravity*— temporal gravity.

### Biological Energy

Biological energy, or *bioenergy*[232] is the "Level 2" version of electrical, magnetic and gravitational forces, where these atomic properties are intermixed with their conjugates—the temporal, "antimatter" version of force and energy that comprises the invisible, cosmic half of the life unit, the anima. What this interaction does is to create an energetic *aura* around *both* the spatial corpus and the temporal anima—what is termed the *etheric body*.[233]

*Force and Energy*

### The Aura of Life Units

---

[232] *Bioenergy* in the sense of the various forms of *biological energy*, not the common reference to energy obtained from biomass, such as biodiesel fuel.

[233] The ether (or æther), in the context of the Reciprocal System, is the cosmic sector of 3D time being interpreted from a material sector, 3D spatial perspective.

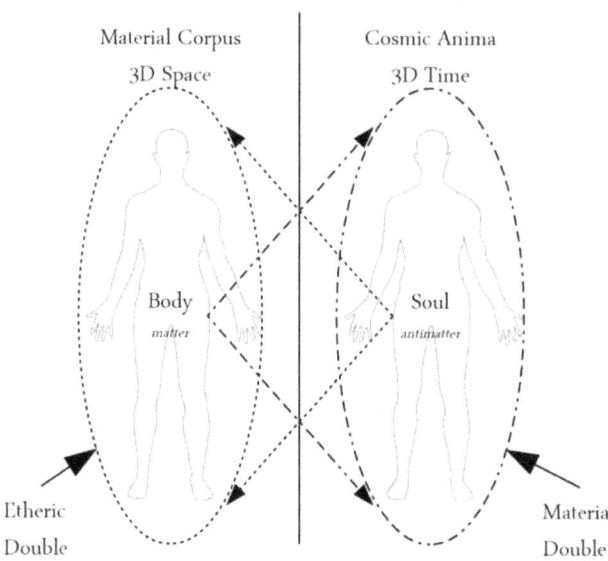

The unobservable half of the life unit, existing in 3D time, casts its shadow around the life unit for the same cause and reason that electric and magnetic fields cast their shadow about inanimate structures—except in the case of the life unit, they are referred to as *auras*. These auras form the *etheric double* of an organism, which is the shadow of the anima about the corpus, the body in space. All life has an aura, and the aura reflects the condition of the cosmic half of the life unit, its "soul." The soul is the seat of emotion, so the etheric double tends to reflect the emotional state of the life unit.

A second auric field, not generally advertised Unit Speed by mystics,[234] also forms around the *cosmic* anima creating an analogous structure to the etheric body that surrounds the physical body. This *material double* shows the condition of the *physical body* to the intuitive side of life, which is how a person can perceive that someone "feels sick" (as mothers always seem to know about their children, regardless of how far away they are). Being a nonlocal connection, it works analogously to the Einstein-Podolsky-Rosen bridge[235] because the souls can remain *adjacent in time*, as a type of "soul group" for family and friends, while the body may be separated by significant spatial distances.

### Forms of Bioenergy

The life unit can be considered a "more complete" expression of scalar motion because it contains *both sectors* in a single, cellular structure. Atoms and particles, being *either* spatial or temporal, only have "half the story," so to speak. Whereas the universe is defined by three dimensions of motion, it is only natural that we would find three expressions of bioenergy in living organisms, reflecting the three

---

[234] This *material double* is the mechanism used to inflict bodily harm in certain martial arts, by energetically attacking the projection of the material body in time. That energy damage is reflected in the physical body after an interval of clock time, giving rise to the famed "touch of death" made famous by practices such as Dim Mak.

[235] The classic *wormhole* that connects things remotely across space, though there is no actual need for a connecting tunnel, as the linkage is just the inverse relationship of the ratio of motion.

*speed ranges* associated with astronomical motion.[236]

Every culture has its own names for these three forms of bioenergy. For the purposes of this analysis, the Chinese terms will be used:

- *jing* (fohat), cellular, libido or sexual energy, associated with health and the corpus.
- *qi* (ch'i or prâna), life force, associated with the anima.
- *shen* (kundalini), consciousness, spiritual and intellectual energy, associated with the animus.

We all know that to store energy you need some kind of battery, and complex biological systems come with such batteries— known in Chinese as the *dan tien* or the alchemical *elixir field* (the *elixir*, itself, is another term for bioenergy).

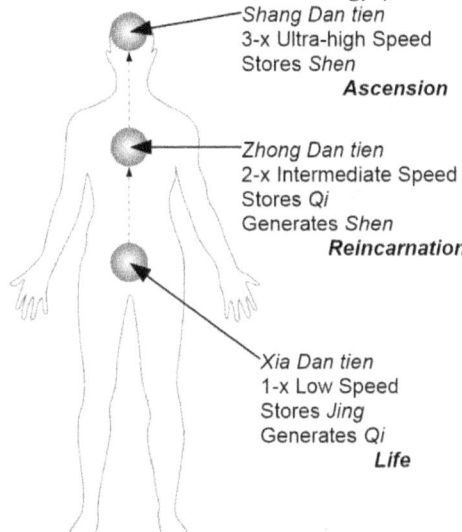

Shang Dan tien
3-x Ultra-high Speed
Stores *Shen*
**Ascension**

Zhong Dan tien
2-x Intermediate Speed
Stores *Qi*
Generates *Shen*
**Reincarnation**

Xia Dan tien
1-x Low Speed
Stores *Jing*
Generates *Qi*
**Life**

These structures, being rotational in nature, exist in an analog to Larson's *equivalent space*, though more aptly named *equivalent motion* for a life unit, as the equivalency includes the merging of equivalent space with *equivalent time*. These equivalent motion structures are not directly observable from our conventional reference system, but are localized at three areas in the body, as indicated, and can strongly effect both the physical corpus and temporal anima. They operate much like a charged electron, where the *dan tien* is the electron, putting out "lines of force," called *meridians*, throughout the body to collect and distribute bioenergy.

## Jing: Cellular Energy

Jing is analogous to Larson's "low speed" range (1-x), being a binding energy within and between cells. Jing is acquired through consumption of food—other living cells that emit jing as they are broken down for use in the body. The better quality the food (such as organic), the more abundant the jing is. Very poor food[237] may have so little jing that the body, itself, must supply a portion of it to aid in

---

[236] Stellar events and life units actually have a lot in common with the way they interact at faster-than-light velocities. One can almost consider a cell to be a thermally-balanced, microscopic star.

[237] This includes genetically modified organisms, poorly fertilized plants and animals pumped full of hormones to produce physical muscle, but not the corresponding levels of bioenergy. The best beef is "free range" because of this—the cattle are *using* their muscles, grazing many miles in a day, which produces a proper balance of structure to energy.

digestion.[238]

Jing is also provided during *conception*, where the new life is provided with a quantity of *congenital jing* from the parents to initialize the life system for independent operation. When the baby is in the womb, it is dependent upon the activities of the mother to do the initial charge of these bioenergy batteries, as those "jumper cables" are only disconnected upon the severing of the umbilical cord at birth.

After birth, natural processes work to acquire additional jing from outside sources (*acquired jing*). Jing is expended with day-to-day activities and consumed quickly in stressful and fearful situations, these days known as "everyday life."

When the total quantity of jing is expended at a biological organism level,[239] the natural process of *dying* is initiated. If the organism does not have sufficient qi to engage the reincarnation process, the result is *death*—a *demotion*. For most of the lower forms of life that are still bound to a collective anima, this demotion is a return to the inanimate, chemical realm. Higher forms of life, having more complexity, get demoted to a level of complexity that is maintainable with their remaining, bioenergy level.[240]

*Force and Energy*

## Qi: Life Energy

Jing is the low-speed energy that keeps cells on *either* the material or cosmic side functioning. Qi is the next step up, the "intermediate speed" energy (2-x) that keeps the material and cosmic aggregates together and communicating.[241] Because this stable matter-antimatter bond is what defines life, qi is considered *life energy*.

Qi can be manufactured from jing or obtained from the environment. Because jing is far more abundant and very easily obtained from a good meal, qi is normally manufactured from jing through the Xia dan tien, then sent for storage in the Zhong dan tien.

Environmental qi is actually quite abundant in natural areas and many people can sense this with a good feeling when out hiking or camping. The primary source is good, clean air, which is why qi is associated with breath, and why practices to encourage and store qi concern breathing exercises. Those that practice these exercises in remote areas often notice that it is far more effective when there are thunderstorms in the area. There is a reason for this: the intense, dielectric field produced by a thunderstorm alters the form of water vapor slightly, converting

---

[238] That "loss of energy" feeling you sometimes get after eating a meal is often due to the body having to supply the bioenergy to digest it, rather than it being part of the food being eaten.

[239] Cells can exchange qi with jing to assist in keeping an organism alive. The process is known as *healing*.

[240] This is the origin of *caste* systems, normally regulated by karma. This demotion process reverses the evolutionary pattern of complexity, returning the life unit aggregate to a simpler level to try again.

[241] Known in some esoteric circles as the "silver cord" that connects the body and soul.

inanimate water into *living water*,[242] a direct source of life units having a structure easily converted to qi.

Unfortunately, few of these places exist in civilized countries that have been overrun by electromagnetism, as this form of energy prevents environmental qi from forming in any abundance.

### Shen: Spirit Energy

Qi can be converted to shen as an *act of will*. Shen is not normally acquired from the environment; it is only obtained as a deliberate, conscious act (however, the process of reincarnation will provide a quantity of shen to the newborn, as well as some shen obtained from the parents). When a life form obtains sufficient shen, it activates the spirit complex—the *intelligence*—and develops a degree of ethics, as Larson describes in his "Level 3: ethical" approach in *Beyond Space and Time*.

Shen can be considered *intellectual* or *creative* energy, depending upon whether the material corpus or cosmic anima is dominant in the organism. Those on a typical, alchemical path that are trying to balance the masculine and feminine aspects within, will tend to exhibit *both* forms of shen, being intellectually creative (composers) and creatively intellectual (inventors).

With the exception of the *Gaia*[243] hypothesis, Nature has not yet developed sufficiently to be an environmental source of shen, so any external accumulation is done from creative and intellectual pursuits involving others of similar mind and spirit; a process analogous to resonance amplification.

When jing runs out and there is sufficient shen, another option for death opens up—that which is commonly called *ascension*, a process to exit the cycle of life, death and reincarnation, by moving to a different realm. If one does not have sufficient shen to ascend, it most likely *will* have sufficient qi to reincarnate and follow that route to try again.

*"When the mind is enlightened, the spirit is freed; the body matters not."*[244]

### Yin-Yang = Time-Space

Another concept that is used from the Chinese is that of *yin-yang*, the concept of "inseparable opposites" that Larson calls *motion*. Yin is the involutive concept: feminine, polar, curved, full or cold, whereas yang is the evolutive concept: masculine, linear, straight, empty or hot. Within the material sector context of the *Reciprocal System* used in this analysis, *time* is yin and *space* is yang.[245]

Yin-yang also represents a way to represent the concept of *simple harmonic motion*

---

[242] *Living water*, as documented in detail by Viktor Schauberger, has the atomic structure of *antihydrogen hydroxide*—one of the hydrogen atoms of the water gets accelerated to superluminal velocities, pushing it over to the cosmic side, converting water into the structure of a *life unit*. This is also the source of *Brown's Gas*, aka *oxyhydrogen*.

[243] *Gaia* being the collective consciousness of the planet and possessing intelligence from that consciousness.

[244] Oma Desala, *Stargate SG-1*, episode "Meridian," and the monk at Kheb on "Maternal Instinct."

[245] Yin-yang and time-space represent the *same* concept. Physicists just don't like those "metaphysical" terms in science!

(SHM), where a yin (inside) SHM is a *vibration*, and a yang (outside) SHM is an *oscillation*.

Vibration, when applied to a rotation such as an atomic system, creates the concept of *charge*, a *rotational vibration* in the *Reciprocal System* that is the source of electric, magnetic and gravitational charges.[246] Charge is *energy*, and the charge on an electron is expressed as *force*—the motion the charge causes on atomic systems. Charges are easily *acquired*, but not *required*. Particles and atoms can exist in either a *charged* or *uncharged* state, a concept not recognized by conventional physics. Because the field effects of an uncharged particle do not vary with respect to time or space, they are not readily detected. The most common example of an uncharged structure is that of the uncharged electron, which is observed as a "hole," the positive "charge" of electric current. The charged electron is the conventional electron, acting as static electricity.

All the forms of bioenergy, like their inanimate counterparts, have both a polarity (poles) and can exist in a *charged* or *uncharged* state. Each type of bioenergy can exist in *four* different states, uncharged yin, charged yin, uncharged yang or charged yang, resulting in what appears to be *twelve* different forms of bioenergy: four types of jing, four types of qi and four types of shen.

**Death and Dying**

The final concept that needs clarification right up front, is that of *death*. The term "death" can be viewed as both a *condition* and a *process*. In this Paper, the word "death" will be used to represent the terminal *condition* and "dying" to represent the *process*.

- *Dying*: the process initiated after the biological organism cannot retain viability.

- *Death*: the condition where the organism gets demoted or returned to inanimate status.

The process of dying is initiated either *naturally* (old age), *involuntarily* (disease or injury) or *unnaturally* (murder, accident). The bioenergetic condition is different for each process:

- *Natural*: The bioenergy of jing has been depleted. Qi and shen are unaffected.

- *Involuntary*: Stress depletes jing by converting it to qi to operate the immune response, until jing is expended. Fighting the disease and regenerating tissue will expend qi.

- *Unnatural*: Remaining jing is converted to qi in large quantities, which may cause the etheric double to obtain sufficient cohesion to take on a "life" of its own. But without a viable, spatial body to contain it, the etheric double becomes a *ghost*.[247]

*Origins of the Afterlife Choices*

**Origins of the Afterlife Choices**

---

[246] Conventional physics only recognizes electric charge; magnetic charge is treated as momentum and gravitational charge as isotopic mass.

[247] "Ghost" will be used as the shadow of the cosmic aspect of life units, not "spirit," which is a common translation for the animus.

The majority of western concepts concerning the afterlife can be traced back to a single source, Plato's *Myth of Er*,[248] which goes like this:

With many other souls as his companions, Er had come across an awesome place with four openings—two into and out of the sky and two into and out of the earth. Judges sat between these openings and ordered the souls which path to follow: the good were guided into the path in the sky, the immoral were directed below. But when Er approached the judges he was told to remain, listening and observing in order to report his experience to mankind.

Meanwhile from the other opening in the sky, clean souls floated down, recounting beautiful sights and wondrous feelings. Others, returning from the earth, appeared dirty, haggard and tired, crying in despair when recounting their awful experience, as each was required to pay a tenfold penalty for all the wicked deeds committed when alive. There were some, however, that could not be released from the underground. Murderers, tyrants and other non-political criminals were doomed to remain by the exit of the underground, unable to escape.

After seven days in the meadow the souls and Er were required to travel further. After four days they reached a place where they could see a rainbow shaft of light brighter than any they had seen before. After another day's travel they reached it. This was the spindle of *Necessity*. Several women, including Lady Necessity, her daughters and the Sirens were present. The souls were then organized into rows and were each given a lottery token apart from Er.

Then of their lottery tokens, they were required to come forward in order and choose their next life. Er recalled the first to choose a new life, a man who had not known the terrors of the underground, but had been rewarded in the sky, hastily chose a powerful dictatorship. Upon further inspection he realized that, among other atrocities, he was destined to eat his own children. Er observed that this was often the case of those who had been through the path in the sky, whereas those who had been punished often chose a better life. Many preferred a life different from their previous experience. Animals chose human lives while humans often chose the apparently easier lives of animals.

After this each soul was assigned a guardian spirit to help them through their life. They passed under the throne of Lady Necessity, then traveled to the Plane of Oblivion, where the River of Forgetfulness (River Lethe) flowed. Each soul was required to drink some of the water, in varying quantities, apart from Er. As they drank, each soul forgot everything. As they lay down at night to sleep each soul was lifted up into the night in various directions for rebirth, completing their journey. Er remembered nothing of the journey back to his body. He opened his eyes to find himself lying on the funeral pyre, early in the morning, and able to recall his journey through the afterlife.

As you can see for yourself, all the classic elements are here. The path in the sky became Heaven, the path in the Earth became Hell, the idea of reincarnation and the veil of forgetting… even those "cast out" of heaven through that 2nd door to

---

[248] *The Republic*, Plato, 380 BCE, 10.614-10.621; translation taken from the Wikipedia entry Myth_of_Er.

take on lives of atrocities.

If we make use of the Annunaki context that is presented in this Series, what we have here is the classic dichotomy of the battle between Enlil (sky god) and Enki (earth god) and the promises made to their slave population, humanity. Enlil (and his father, An), lived in the "heavens" and would come down to the surface world as rulers of the land. Enki, with his undersea Abzu and handling mining operations across the planet, lived beneath the surface and would come up to the surface world, bring the treasures from below to placate the sky dwellers. With the onset of hostilities between them, these domains of control became the classic Heaven (Enlil as Yahweh/Jehovah) and Underworld (Enki, the Adversary[249]). The role of human souls in this Annunaki scenario will be expounded upon later, after some foundational concepts have been reviewed.

**Options While Dying**

The quantity and type of bioenergy acquired during life provides options when dying of natural or involuntary causes. Unnatural death causes a situation where the remaining jing (the years taken from you, so to speak) are converted directly to qi upon the loss of the corpus. As a consequence to this boost in qi, *death* from unnatural causes seldom occurs, as the boost to qi jumps an entity directly to the reincarnation option. This is a type of "fail-safe" that is embedded into the system to give those entities a second chance, providing they don't want to remain earthbound to spook others as ghosts.

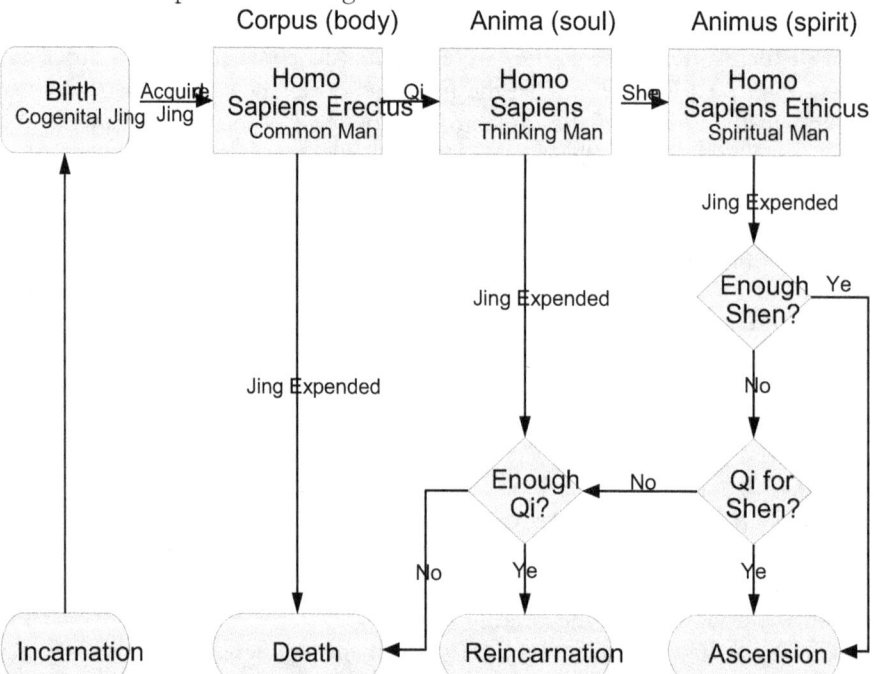

Reincarnation and ascension are entry points into a more complex system of tests

---

[249] Enki, having accused Enlil of working against the best interests of the Annunaki, became Enlil's biggest adversary and was given the title of the "accusing adversary," *satan*. "Satan" is a *title*, not a person, much like "President."

and consequences, where concepts such as *karma*[250] come into play as well as evolutionary history. Here, the concept of death can be considered analogous to "flunking out" in school, in which you return to a prior grade to learn the lessons you obviously missed, and try again. True death, a total return to inanimate status, is basically flunking out of Kindergarten, so it is infrequent.

**Immortality**

A natural consequence of this structure is that if life were to acquire jing at a rate faster than it were to expend jing, the natural process of dying would *never* occur and one could live forever. That is true, but there is a *catch*, which has to do with our old friend, the *magnetic ionization level*[251] that is responsible for determining which elements are radioactive.

A non-zero magnetic ionization level allows atoms to collect neutrinos to add isotopic mass. In the inanimate realm, atoms can remove this excess mass through radioactivity, but in the case of the life

*Immortality*

unit where the material and cosmic structures are in balance, isotopic mass builds on *both* sides. The material and cosmic structures continually increase their atomic displacements until the isotopic mass reaches a point where it neutralizes the atomic rotation.[252] This causes the linkage between the material and cosmic structures of the life unit to disassociate, causing *cell death*. By acquiring more jing, you can extend your lifespan, but you will continue to "get old."[253]

But as ancient history teaches, the "gods" were immortal, so they found a way—but not by extending the life of a single, physical body—but by creating bodies "on demand" and transferring their personality from one body to another, through the use of what was termed, "black magick." There is a way to achieve immortality for humans, as documented by the ancient, Taoist masters—if you are willing to pay the price. Curiously, many of the Taoist masters that achieved immortality and departed the mortal realm for the realm of the gods, often *returned* to live out mortal lives—though never explaining why.[254]

**The Death Experience**

If you want to understand death, the last place you want to look is modern religion. And the second to last place to look is to the "New Age" experts. Go to the source, which starts with the Sumerian records and the books written by the

---

[250] Karma is the carry-over of actions and intentions from prior incarnations that influence the current one.

[251] In the *Reciprocal System*, the magnetic ionization level, an environmental variable, controls the amount of isotopic mass an atom can accumulate before undergoing radioactive decay to eliminate the excess mass.

[252] One would think that as one ages their weight would increase, due to the additional isotopic mass—as the bathroom scale seems to demonstrate to me, every day. But that is not the case; because mass and anti-mass is being added together, there is no net observable change in atomic mass within the cell.

[253] Wherever Nature provides a problem, it also provides a solution... such as the *Fountain of Youth*, which may not be the fairytale people think it is.

[254] I've wondered what "heavenly" situation could be so disruptive, as to make one give up immortality? Perhaps it's those 24/7 "Elvis" concerts!

gods of old, such as the *Book of Jubilees*, which is a dictation from God recorded by Moses on Mount Sinai, while He etched those stone tablets for the Israelites. Yes, God *did create* man—though the genetic engineering context got a bit lost in the translation. So if anyone understands the structure of mankind, it's the folks that designed the hybrid—and that information is good reference material.

What is described are two *realms*, the *mortal realm* of our everyday existence and this "other" realm, which I'll just refer to as the *Other Realm*,[255] as there are just too many names in use, all with extreme religious connotations. If you've done your homework and followed up on Larson's *Reciprocal System*, you already know what these two realms are: the two "sector" aspects of the biological level of existence, namely a "2nd density" version of the material and cosmic sectors.[256]

When it comes to dying, humans, the Cro-Magnon descendants, are a mixed bag—a hybrid of both natural processes and "divine" processes from their Saurian creators. The bulk of humanity is more the "down to earth" type, consisting of a genetic mix of Cro-Magnon and Neanderthal—the "red blooded" folks. A smaller subset, billing themselves as the "blue bloods" of Royal descent, operate more on the Saurian "top, down" principles, than those of the evolutionary natives. The natural process will be addressed here, as natural consequences of life.

Death is a deconstruction of the biological level of existence, which may reach all the way back to the inanimate level. Like any radiative or radioactive process, energy is released, and in the life unit

---

[255] Concept borrowed from the television series, *Sabrina, the Teenage Witch*, where the non-mortals (witches, wizards, mythical beings, etc.) can exist, and move between, the mortal and "other" realms.

[256] A "realm" is a better term than "sector" in this situation, as a *sector* is from *section*, a piece. The biological realm is *both* material plus cosmic, so it is inclusive of both sectors.

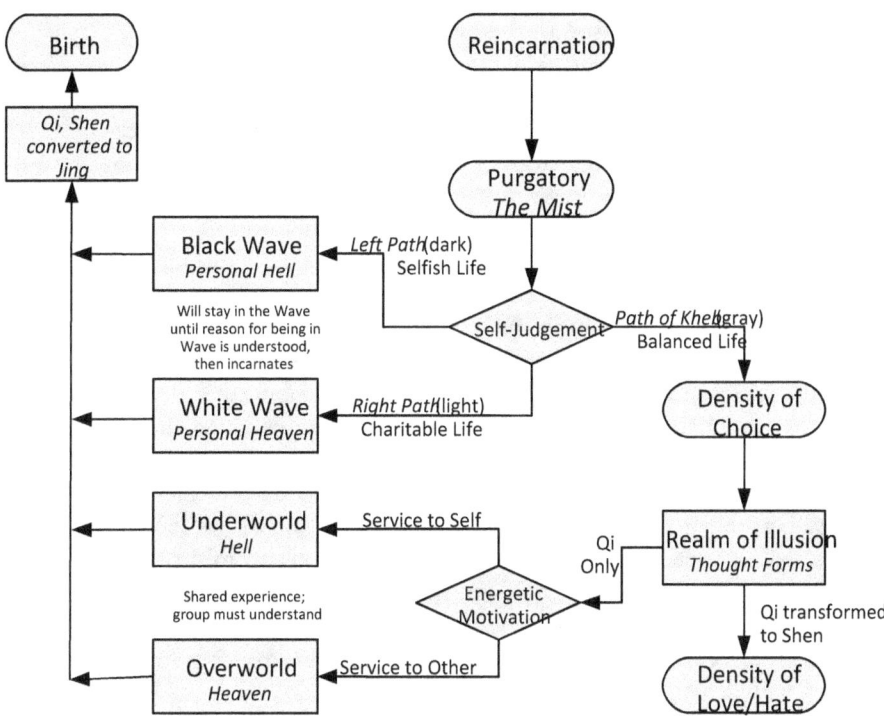

situation, *bioenergy* is liberated. But unlike electrons, alpha particles and gamma rays, bioenergy continues to have a *continuity of form* for a period of clock time, still able to slightly influence the spatial realm through the forces generated by the anima. The higher functions of consciousness, being nonphysical, remain intact and operating, leaving this etheric ghost in a transitional state. Post-death rituals, such as those documented in the *Bardo Thodol*, the Tibetan *Book of the Dead*, were created to assist this ghost in the options available to them, during this period.

Unnatural death has the tendency to "explode" the etheric body out of the physical, sometimes to such a degree that a person does not even know that they have lost their body. Due to the high level of qi, and the fact they don't realize something happened, no attempt is made to "move on" and the ghost remains earthbound until either such realization happens and they can make their choice, or qi (which cannot be replenished) dissipates to a point where a choice is forced.

**The Reincarnation Process**

The presence of qi allows a secondary process to activate when dying—that of *reincarnation*. If qi is totally inadequate (less than *one* natural unit, as Larson would put it), then there is not enough to power the afterlife system, so it is back to inanimate status. Higher forms of life tend to have accumulated a large amount of qi, so the reincarnation process is normally available.

People have been through the reincarnation cycle a considerable number of times over the last five millennia, and when born, already have a quantity of qi and shen carried over from the previous cycle. Qi often determines the relative health of the newborn, whereas shen is the determining factor for karma. As in all energetic systems, bioenergy has both *charge* and *polarity*. A person with bad qi may end up deformed upon reincarnation, whereas a person with good qi may end up athletic.

Good shen will give a person a more virtuous life, whereas bad shen may cause constant problems and a fearful life.

*The Reincarnation Process*

In order to understand *reincarnation*, one must have some idea of what "incarnation" is. It is derived from "carnate," which means "to be put in flesh" and that means to *come alive*. *Reincarnation* is to become alive—again—inferring that there was something that carried over from a prior incarnation.

Upon death, the material and cosmic atoms disassociate, returning the corpus to inanimate status. But the bioenergy field remains as the ghost or etheric double. It is in this state that the consciousness experiences the afterlife situation.

In the diagram above, I have identified the first afterlife realm as "Purgatory," an old Catholic term for a place that souls reside immediately after death to do a life review to see where they'll end up for eternity. In researching this realm, most systems add an additional step, what can be called "The Mist."

When consciousness loses the physical senses as the body dies, it gets a bit confused, as things like *sight* are only available from what is known as the *3rd eye*. Things start out rather hazy at first, like looking through a mist or dense fog. At this time, the consciousness is still very connected with the spatial environment—people, structures and reference points can still be identified, just not as physical systems—only energetic ones. Without a body, you haven't got a "ghost of a chance" with actual, physical contact.

The dying experience continues in this state for 40-60 days, depending on the culture.[257] It is thought that during this time, the folks one has left behind have the opportunity to still communicate with you to say their final thoughts and get closure, before you head on to more interesting things. Also remember that movement, in this ghostly form, is not spatial—it is *energetic*, so your consciousness can localize at any point on the Earth when it needs to. And that need usually arises when a loved one is thinking about you. So it does seem probable that this period of transition exists to get closure.

After interactions with the physical world are complete, you have gotten a pretty good idea of what you were like as a person and enter a process of "life review." This is where *you* decide what is going to happen next, and one can be quite hard on themselves. In our society the victim status is rewarded, which leads to a sense of worthlessness that *does* carry into the afterlife.

In the classic religions, the life review ends up with two, simple choices: *selfish* and "go to Hell" (the black or dark aspect of the wave) or *charitable* and go to Heaven (the white or light aspect). Note that these "waves," like their oceanic equivalent, are *transient*—a *personalized* version of heaven or hell, not the collective version that was initially established by the gods of old.

The mind is a powerful thing, and since the mind is the cosmic aspect of life (with the brain being the material aspect), it survives death of the body but is no longer constrained by the shared, spatial illusion we call "reality." Your mind can create any illusion it desires in order to balance out your bioenergy, so you may continue on the path of spiritual evolution.

---

[257] *Mo Pai*, 40 days; *Bardo*, 49 days; *Native American*, 2 moons ~56 days.

Those having a near-death experience typically enter the waves, just like a person on the shore wading into the splashing of the ocean. They have not fully engaged the dying process because at some level they knew they were not going to die. But, they do get a glimpse of what personal judgment may result in, and as such, the experience can often make a *substantial* change to a person's outlook on life—normally for the better—once they discover that a personal hell is built upon all those things that they spent their life repressing and hiding from.

There is also a third choice, which is labeled "The Path of Kheb."[258] This path opens up when, during life review, when you realize that "yes, I have done some selfish things, but they were necessary for survival and growth," concurrent with "yes, I have done charitable works, and they were also necessary to balance out the selfish acts." Most people that are not strongly biased in their political, religious and philosophical views fit this description. When the consciousness enters this state, neither a personalized heaven nor hell suffices to provide further growth of consciousness, so "Door #3"[259] opens up—as the Minbari[260] say, "*I am gray. I stand between the candle and the star. We are gray. We stand between the darkness and the light.*" This removes one from the personal realm of self-judgment, the "I," and into the *Density of Choice*,[261] the "We," and what interactions—selfish or charitable—have been done in order to assist in the evolution of consciousness.

### The Realm of Illusion: The Density of Choice

An integral part of the transition zone particularly applicable to the more complex organisms such as humanity, is the *Realm of Illusion*, referred to in the *Law of One* material as the *Density of Choice*. It is reached when there is sufficient qi and one has judged their life to have been worthwhile; they have accomplished what they set out to do in life and are at relative peace with the complexes of the psyche.

The *Realm of Illusion* considers the larger picture—not only what you did with your life, but how your actions have affected others, with a single determining factor: did your actions help the *evolution of consciousness*? This determination is non-judgmental as to orientation, as a self-serving act may result in the evolution of consciousness, just as easily as a charitable act would.

This realm is described in ancient records as containing an environment that is manifested by the *intentions* behind the actions you took during your lifetime (or deathtime, as the case may be), as the immortal realms consider *intention* over *action*. It is a realm of thought-forms, created by the qi bioenergy. When one chooses to treat this artificial reality as a real, tangible environment, they have elected to reincarnate through one of the service paths: *service to self* or *service to other*. The qi of the person is released into the environment, so they can experience this service

---

[258] From an Egyptian story and also the references to Kheb in the *Stargate SG-1* television series, as a place to learn the path of ascension.

[259] A comical reference to an old television show, "Let's Make a Deal," where contestants were given options to keep prizes they had won, or to risk them and to pick a door, something behind a curtain, or the box Carol was standing next to.

[260] The Minbari are a philosophical, alien species in the television series, *Babylon 5*.

[261] *The Ra Material* lists seven densities of existence, or levels of complexity. The densities are used in this work for lack of better terms.

path and eventually be born again, in the mortal realm or the other realm, depending on the one they departed.

Here, entities can remain for centuries of clock time, until they figure out that they are basically on a holodeck[262] of their own making and the environment they are living in is nothing more than a projection of their own unconscious. That realization will stir the movement of the bioenergy field that has been used to create all the structure and interactions around you. Your artificial reality collapses and you get an influx of qi—so what do you do with it? At this point, what was *real* becomes *unreal*[263] and the qi that was sustaining the system is returned to conscious control—but at an elevated stage: the energy returns as shen, indicating that the reincarnation process is not going to work for the further development of the consciousness of the organism.

Some will get "pissed off" and use that qi to repel the world around them. This normally happens with *service to self* based individuals, whom believe that they are the center of the universe to begin with. As a result, balance is lost and they proceed to the *Underworld*, the more classic Christian Hell, where this energy can be literally "burned off."

*The Reincarnation Process*

Alternately, some will fall in love with the experience, handing their qi off to others and get sent to the *Overworld*, where they can continue the experience for a very long and loving time—until they run out of qi and fall back into the incarnative process.

Again, there is a 3rd path available that emerges after you realize you're on the holodeck and accept this artificial reality of thought forms for what it is—a *really good education on how to balance yourself*. At that point, you can just "call up the Arch" and "Computer, end program."[264] The released qi has no where to go via an unconscious act of *service*, so it is *yours* to command. After all, qi *is* intelligent. So introspect on your holodeck experience, and that introspection will convert qi to shen and open up the 3rd path, the first stage of *Ascension*.

---

[262] A "holodeck" is an artificial reality suite, popularized in the *Star Trek, the Next Generation* television series.

[263] A concept played upon heavily in "The Matrix" films, concerning the red and blue pills.

[264] The holodeck contains an archway with a computer interface, that can be called into manifestation during a simulation in order to alter or end the running program.

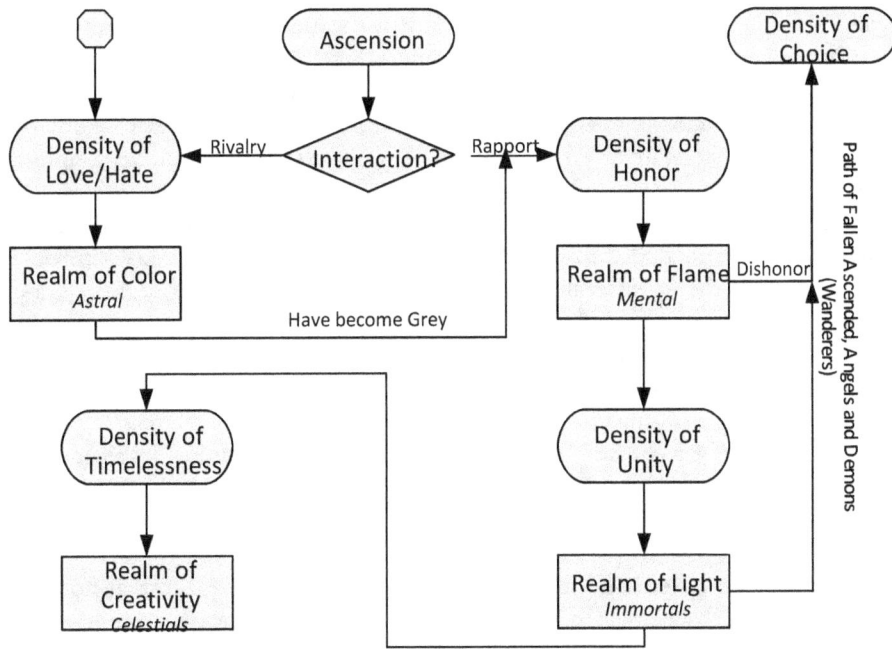

**The Ascension Process**

When the ethical man, through a conscious and knowing act, generates sufficient shen prior to expending his jing, the option becomes available to skip the whole reincarnation drama. There are two entry points into this realm, either through the *Density of Love* (or *Hate*, 4th density) or through the *Density of Honor* (5th density). The selection depends on how you generated the shen—through *rivalry* or *rapport*.

*The Realm of Color: The Density of Love and Hate*

There are two doors that enter the realm of color, one from the *Realm of Illusion* and the other directly from the process of dying with a sufficiency of shen, but still engaged in the discord of *rivalry* or *competition*. Entry into this realm indicates that the organism has developed a sufficient animus (with both thinking and feeling aspects) to exit the reincarnative system, but has not yet reached the status of immortality, which begins with entry into the *Density of Honor*.

This *Density of Love and Hate*[265] (not "like and dislike"), the 4th density of the *Law of One*, is a colorful place where the personal choices are at extremes. It is known as the *Realm of Color* because of the extreme nature of the values and choices made. Many people that go on chemically-induced, psychedelic experiences end up here for a short time.[266] Looks good when you get there, but one does not realize what they are actually experiencing because they have skipped over the process to arrive at that destination. This realm acts like an overlay to *both* the material sector and the cosmic sector, sitting beyond space and time. From this vantage point, both

---

[265] The *Law of One* material omits the "and Hate" part, though the dichotomy is obvious from the descriptions.

[266] Hence the predominance of bright colors and patterns preferred by psychoactive drug users.

sectors can be viewed and interacted with at the same time; the concept of past and future changes into a complex landscape, where a person's body can be seen following one path through *space* and their soul can be seen following another path through *time*. Much of the metaphysical folklore regarding travelers from the future or past originate from this realm, as pasts and futures are geographical directions in time.

As there are two entrances, there are also two exits from this realm. The first is when shen is expended through continued rivalry and discord, usually through interference in the lower realms. The second exit opens when one brings into balance the extremes of color and becomes *gray*... a system in sympathy and rapport, which leads to the immortal aspect of ascension.

The use of rapport and rivalry do not indicate a dichotomy at this level, but more a level of complexification or dissolution of consciousness. Shen is a *conscious* energy, whereas qi is an *intelligent* energy. Those who engaged in rivalry to generate shen have basically fought the system using the attractive and repulsive principles of love and hate. Competition can be friendly or hostile, and that forms the basis of rivalry.

Rapport, on the other hand, is the "pro-consciousness" approach that not only provided yourself with shen, but assisted others in getting it along the way (frequently confused with *service to others*). When rapport is the basis of shen, then one needs not go through the resolution of the competitive methodology and can proceed to the *Density of Honor*, manifest in the *Realm of Flame*.[267] The realm is analogous to the "mental" plane of other systems, where one obtains the ability to create and destroy structures in the physical, spatio-temporal worlds.

The reference the *Realm of Flame* comes from descriptions of the energetic body resembling that of a flame—yellow or golden in color, vibrant, energetic, though still anthropomorphic. Yellow is the color of mental energy, which is why this realm is equated with the mental plane.

**Overview of the Mortal and Immortal Realms**

The *mortal realms* are the realms in which we have physical existence. In the *Reciprocal System*, physical can be either spatial or temporal, so the mortal realms are *either* existence in the material sector of 3D space or the cosmic sector of 3D time. When your consciousness is in one sector, the other sector has the appearance of the "afterlife." Most theological conventions consider where your consciousness currently resides is the *mortal realm* and its conjugate, the *afterlife*, is the *other realm*.

What many folks do not realize is that even when you die, you're still "mortal"— just mortal in the afterlife realm and will, after a deathtime of experiences, get old, be born and come back to life, here. It is a symmetry of existence between the matter of 3D space and the ether of 3D time. The mortal and afterlife realms are just opposite faces of the same coin, so when you go around the edge of that coin, *Overview of the Mortal and Immortal Realms*

you die on one side to be born on the other. The primary difference is that physical existence is valued by space (sensation, thinking, distance, clock time) and the

---

[267] Note: "flame," not "fire," as in *Stargate SG-1*'s Ori. Though this is the realm in which one would find Stargate's ascended Ancients.

afterlife existence is valued by time (intuition, feeling, duration, clock space).

In the mortal realm, "the road to hell is paved with good intentions"[268] and one is valued (judged) by their *actions*, not their intentions. The material realm (life) is based on spatial relationships—the purpose of our physical body is to sense and manipulate space. The cosmic realm (afterlife) is based on temporal relationships—the purpose there is to intuit and manipulate time. Both cases require interaction between consciousness and the environment.

The *immortal realms* are reached by the *ascension* process, breaking the cycle of reincarnation. To use the coin analogy, it is very difficult to jump off the face of a coin, but easy to do when you are making the transition around the edge. The most notable difference between the mortal and immortal realms is that the immortal realms exist *beyond* space and time, and as such, the valuing systems that we are accustomed to are very different—they take on a reciprocal form. There is little difference between *thought* and *manifestation*, so valuing is done by a concept similar to *intention*.

However, in the immortal realm, "the path to *mortality* is paved with *actions*." It is from this premise that we get the non-interference directives of higher beings, such as the ascended Ancients of *Stargate SG-1*. Consider: once you have ascended, you did it because you no longer need to experience the physical interactions (sensation and intuition) present in the life and afterlife realms. In other words, you have graduated High School and have started attending College. But… if one continues to have an interest in interacting with the physical realms, then obviously you did not complete a lesson there, which is why you have the motivation to continue interacting with the physical. Once that link is formed it tends to get stronger, as an ascended being can greatly affect the affairs of the mortal realms due to the fact that their thoughts will manifest as structure and action, substance and force. Eventually, they get sucked back down into the reincarnation cycle to figure out, and finish off, the lesson that they failed to learn. So, when one gets "cast out" of the immortal realm, they return to mortal status with some knowledge of higher learning—but that knowledge is often misunderstood, as the mortal mind/brain does not have the mechanisms to comprehend non-corporeal lessons.[269]

### Ascended Laws of Non-Interference

Do interactions between the mortal and immortal realms occur? Yes, frequently. But it is seldom a *physical* interaction. It normally occurs during a *'tween time* (an in-between time, the edge of the coin, high noon, midnight, sunrise, sunset, birthdays, anniversaries, doorways, windows, bridges, shadows… any time a new

---

[268] Saint Bernard of Clairvaux, circa 1150 CE.

[269] In Don Elkins' book, *Secrets of the UFO*, he describes *Wanderers* as high-density entities that made a "choice" to return to Earth and help out. When you include information from other sources, such Asatru and Native American mythology, a slightly different picture emerges—though one can easily see Elkins' perspective derived from it. Wanderers did not voluntarily re-enter the Density of Choice— they "fell off the wagon," so to speak, and "cast themselves out" because they chose to interact and participate in the physical system again, from the ascended realm, and managed to generate karma as a result.

cycle starts, or opposites become connected). And it is done at a *thought* or *conceptual* level that can provide some very insightful information, but is still up to *an individual* to *act* on it, physically, since that is the dictate of mortal existence. It is from this interaction that we get the concept of spirit guides, ascended masters, avatars, guardians, et al.

There are times when the thoughts and intentions of someone in the *mortal* realm pokes through to the *immortal* realm and gets the attention of immortals. It is usually associated with a flash of insight about the nature of one's purpose in life. It does not happen often but when it does, most choose to ignore it. (Remember that the immortal realm is valued by *intention*: if a mortal does not *intend* to follow through by indicating no interest, putting it on hold, or just denying it, the offer of interaction will be withdrawn—it is usually a one-shot deal.)

When an offer comes from the immortal realm, it indicates that the recipient has a quality of character that would allow them to do bigger things than what biological existence would have them do.

Something to note is that an immortal *guide* will **never** tell you to "do" anything nor act on their behalf. They have *no interest* in the mortal realms, but do have an interest in the larger picture—of which you can make a difference, if you choose to do so. If you encounter a non-corporeal entity that wants you to perform or act on their behalf, you probably have another mortal inbetween lives that has used projection to gain access to this side. The immortals simply give you access to an enormous realm of knowledge and experience, and like a librarian, will assist you in finding what you are looking for—but you will still have to read the book and act on the knowledge.[270]

Interaction between the realms is a free-will choice, and hopefully an *informed* choice.

### The Conjugate Realms: This World and the Otherworld

One of the more spectacular conclusions that Dewey Larson made with his reciprocal relation between space and time was the discovery of a sector of the Universe that is the conjugate of the one we are familiar with—literally, the *Universe of Time*, which he termed the *cosmic sector*. This cosmic half of the universe contains structures similar to what we observe in the material, spatial half: atoms, dust, rock, asteroids, planets, suns, galaxies,… everything that can exist here in 3-dimensional space, can have an analog in 3-dimensional time; these two halves of the universe are not separate phenomena.

The close coupling of these sectors can easily be identified by the difference between *structure* and *force fields*. Structure is spatial; we can observe it, Birth However, force fields are temporal—unobserved except by the effect they have

---

[270] One big concern about psychoactive drug use is that it artificially pokes one's thoughts into the immortal realm, that would not happen under natural circumstances. Native American and aboriginal people respected this; use of psychoactive substances was restricted to shamans and anyone wishing to seek their use required years of instruction prior to their first dose. It was a very respected tradition. But now the immortal realm is learning that humanity is "crying wolf" with these artificial projections and may soon cease making offers, altogether, to humanity.

*on* structure. A simple
of the universe is through the concept of the complex quantity, where the *real* component is space and the *imaginary* component is time. Note that this
is the case from the material perspective; from the cosmic (antimatter or afterlife) perspective, the situation is reversed: time is real (3-dimensional) and space is imaginary. Perfect compliments of each other.

In the ancient mythologies and parables, this "world" refers to the 3-dimensional, spatial realm of observed structures, being ordered by clock time and invisibly influenced by the "otherworld." The properties and characteristics of this otherworld closely match the natural consequences of a realm of Larson's cosmic sector. Reported journeys into the otherworld equate distance and duration. Should a

*Overview of the Mortal and Immortal Realms*

visitor to the otherworld fail to exit it from *precisely* the same point they entered, upon returned, they would be *displaced in clock time*—arriving before birth or years after death. This is the concept of *clock space*, where *distance* in the otherworld equates to *duration* in this world.

This world is identified as the land of the living, entered by birth and exited by death, but only comprising half of the cycle. The otherworld comprises the other half of the cycle, the "afterlife" as it is oft called, entered by death and exited by birth. There are those in this world that do not believe in "life after death." And usually the same folks, after transitioning to the otherworld, do not believe in "life before death." (This usually indicates that they haven't been around the loop very many times.)

## The Transition Zone

As mentioned earlier, if an organism has not acquired sufficient amount of qi upon the dissolution of life units while dying, both "halves" of the organism can return to inanimate status—the dust from which you were created. This is a "true death," in the sense that nothing of the pattern of memory or spirit of the organism is able to be carried forward. It is not a waste, however, as the atoms release in both the material and cosmic sectors will eventually be reused in another living organism. The universe is "recycle friendly."

However, when there is sufficient qi to retain the pattern of identity, the organism enters the transition zone and enters the reincarnative or ascension process. The transitional zone starts with "reincarnation" or "ascension" blocks on the chart and ends with either rebirth (reentering the conjugate sector through birth) or ascension to the immortal realms. Note that the transition zone is *not* the cosmic otherworld. The basic nature of the otherworld is indistinguishable from this world to the newborn consciousness— you believe you have been reincarnated into the same world you left, but have actually transitioned to the conjugate aspect. Upon death in that world, you will again enter the transition zone and return to the land of the living. Though in both cases, "living" and "afterlife" are relative terms.

## The Immortal Realms

The term "immortal" means, "not subject to death, everlasting." In the context of the ascended being, this is indeed true as they have moved beyond the necessity of the reincarnative cycle, as well as the concepts of clock time and clock space.

It is just another stage of existence that is a natural consequence of the evolution of consciousness and does not infer special power or ability, as those concepts are mortal ones, not immortal ones.

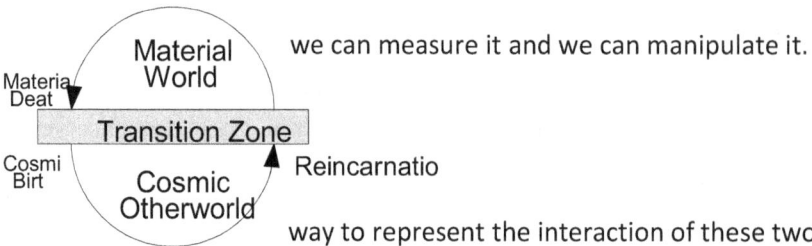

The Realm of Flame: The Density of Honor

The 5th density mentioned in the *Law of One* is the *Density of Honor*, more appropriately called, "duty, honor and responsibility." It is the first of the immortal realms where systems are brought into balance through the concept of *rapport*. Though the system is in balance, it is more like a juggling act than perfect symmetry. The net bioenergy of the organism has reached peace, but it is far from symmetric or crystalline.[271]

Honor has aspects of *duty* and *responsibility*, which play an important role in the *Realm of Flame*, because the primary energy of the realm is *shen*, the animus, intellectual or spiritual bioenergy. The activities here are primarily *mental* ones and this is also known as the *mental plane* of existence, where one has a duty to continue the evolution of consciousness along with a responsibility *not* to interact with the mortal realms. Newly ascended entities are very tempted to interact with the mortal realms, because of the displacements in space and time they carry to hold their balance. But if they do, then that displacement is offset and they literally fall back in to the mortal realm, having not learned their lesson.

Evolution in the Realm of Flame is achieved by burning off the remains of the material and cosmic values, those of self-centeredness, excess charity, service to self and service to other. When the chaff of mortality is cast off, the system becomes more symmetric and the transition to the *Density of Unity* begins and consciousness enters the *Realm of Light*.

---

[271] When something *crystallizes*, it becomes frozen and unchanging with respect to either clock time, or clock space.

### Densities of Unity and Timelessness; the Realms of Light and Creativity

Information from human sources is scarce on these immortal levels of existence, but can be inferred through the extension of Larson's *process of elimination*, by removing what we can attribute to the other densities and realms and taking a look at what is left over. Unfortunately, it does not leave much, other than to indicate that there are two, additional levels that have been achieved by conscious, intelligent beings (hard to call them "human" at this stage of development).

The *Density of Unity* and the *Realm of Light* is the stage at which consciousness takes on a new role, based in rapport and has the ability to be a *creator*, perhaps better termed, *evolver*. It is not about genetically engineering new races or zapping new worlds in or out of existence, but concerns an investigation into the *evolution of consciousness*, itself.

Once this investigation has reached some conclusions, the *Density of Timelessness* and the *Realm of Creativity* become the next step, where these conclusions can be put into action. The research of George Hunt-Williamson on the *Cyclopeans* and the group he terms the *Elder Race* provides additional details.

Using the corrected geochronology developed in Part 1, the first visitors to this planet were the Cyclopeans during the Paleozoic Era and their presence may have well been responsible for the *Cambrian explosion of life*, documented in the fossil record. Williamson's premise was that these visitors were a spiritually advanced race, seeking the next stage of the evolution of consciousness and were looking for an out-of-the-way place to take the final stages of this more advanced form of ascension, not a movement to a higher density, but a transition to the next *octave* of existence. Our solar system, being on the leading edge of the Sagittarius Dwarf galaxy being pulled into the Milky Way,[272] was at the leading edge of solar system evolution and therefore made a good "monastery" for

*The Immortal Realms*

the completion of this research. The Cyclopeans did not directly interfere with the development of life, as did the Titans and their Annunaki progeny, but the presence of a field of consciousness that intense would undoubtedly accelerate the development of biological diversity in the local environment. So our solar system is a special place, being both on the leading edge of solar system formation and having a collective boost to the life energy of the world.

### The Psychology of Souls

Psychology, "the science of the mind, mental states and the processes of the psyche," is not something taught in grammar school—though it *should* be, as it is a type of "owners manual" to your brain. Of course if you understood psychology at an early age, you could not be easily influenced by things like subliminal

---

[272] Our sun is part of a small, irregular dwarf galaxy being absorbed into the Milky Way. Remember that stars consume matter to move from red giants to the main sequence. The Sagittarius Dwarf, sitting outside the rim of the Milky Way did not have much in the way of matter to consume until it began to enter the galactic disk. At that time, abundant "fuel"

marketing, political propaganda (aka "the News") and might actually reap some significant benefits such as finding good, stable relationships with prospective friends and mates, as you would not need to stumble around with "games" to figure out how relationships work.

Many psychologists have already accepted that some kind of "energy transfer" exists that can be described in psychology with the concepts of *projection*, *identification* and *transference*. These are the bioenergetic "ties that bind" that can influence your decisions, get you to act a specific way, or even take your very life energy from you.

These concepts have a simple, inanimate realm analogy: *force fields*. If you are an iron filing, you will be easily—and invisibly—influenced by a magnetic field, or could have an electrical discharge pass right through you. Fortunately, you are *not* an iron filing, but a biological organism, and are therefore subject to the bioenergy version of force fields, an *aura*, which is a nonlocal effect that is represented by the interaction of the etheric and material doubles—not the corpus or anima (as they are both *physical* aggregates).

The most common of these auric interactions is *identification*. This normally happens between a person and an object that they find some strong *attraction to* or are totally *repulsed by*. The object, itself, is inert in the auric range, but its structure, color or memory[54] resonate with your aura and produce feedback. Positive feedback means you just *have* to have that thing—and cannot live without it. This is what sales and marketing plays upon, to get you to buy all this stuff you never knew you needed, but can't live without.[55] Negative feedback pushes you away from an object and is often used to hide things, by making their appearance grotesque.

*Projection* is similar to identification, in that you voluntarily transfer your bioenergy to someone or something else. In a typical relationship, a partner will transfer their "unseen half" onto the other, in order to see what is going on within themselves. Typically, a man will project his anima (soul) onto his wife and the wife will project her anima onto the husband. This projection works just like the projection screen one watches films on and is a *tool* the psyche uses to help to understand oneself. Unfortunately, the actor/actress on this psychological screen has *free will*, behaves the way they normally behave and does not act out the role that has been projected on them—expectations are not met and the relationship suffers. Once you realize that you are both a projector and a screen, then it

---

became available to advance the stars on the leading edge quickly, developing binaries, trinaries and solar systems well in advance of the rest of the dwarf.

54   Inanimate objects do possess a function analogous to memory, which is a structural arrangement in 3D time that is not visible to the spatial observer.

55   In order to generate this auric feedback, marketing often uses sexual symbolism to *identify* their product as a catalyst to engage in sexual intercourse, which is a very strong biological function in humans.

becomes easier to interface with others, as you can begin to identify your own expectations—not because someone acts them out, but because they *fail* to do so. As long as people are fulfilling each others expectations, relationships are maintained (it's called a *complimentary neurosis*). Once that fulfillment stops, the relationship is terminated and all the bioenergy that is tied up in that exchange gets released with a snap, comes hurtling back into yourself, and in most cases, the result is anger and emotional injury, just because the other person was a terrible actor for the role you gave them to play.

Projection can take on a more serious and potentially life-threatening form, known as *transference*. Rather than just giving someone a role to play on your stage in life, one *removes* that entire chunk of their psyche and hands it to the other person, script and all, and you no longer have any control over it —only they do. And when they fail to meet your expectations, you *do not* get the bioenergy back, as you have literally given them a piece of your soul. They get to keep it and use it for themselves.

Transference is the mechanism used by people that want *power over you*, namely anyone seeking power, control or *fame* in the public eye. And they are fairly easy to identify as they will use *possessive* terms like "my people," "dear ones," "constituents" or the like. They use the tool of charisma to get you to *project* some need you want fulfilled onto *them* to fulfill, basically volunteering to play that role for you. Once you have projected those desires, your auras become locked together—and distance does not matter, as it is a nonlocal, bioenergetic connection. Then they take steps towards that fulfillment, usually by making savior-type promises and even providing some artificial evidence that attempts to lock you into the system (a common, political tactic). You have been nibbling at their bait, and they *want* you to swallow that hook and fully engage in transference, as they normally do not have the ethical development to obtain that energy for themselves. And once they get it, one of two things will happen: first, if you're a good slave to them, you will go out and recruit more bioenergy "food," increasing the size of their contingent of followers. Second, if you have outlived your usefulness, they will say or do something to make you reject them and leave so they can *keep* the bioenergy of the transference. Look at *any* of the "leaders" of the world and you can see this in full operation.

### *Cheating Death*

The psychology of transference, when done at extreme levels, gives an organism the ability to *cheat death* because the choices available during dying are *based* on bioenergy. Bioenergy, being organic in nature, does have a "footprint" from the life form it belongs to, but that footprint is erased and replaced once *voluntary* transference of bioenergy to another organism is made.[273] On the plus side, it allows life to provide life energy to another life for the purpose of health and healing. On the minus side, it can cause life to get "demoted" from lack of bioenergy that was transferred away, when they otherwise may have had a shot at ascension.

Since there are significant karmic repercussions in this "cheating death" scenario, those that engage in the practice desire not only to *avoid death*—but *avoid the entire cycle of reincarnation*, altogether. This can be accomplished using techniques commonly known as "black magick," though it is actually just the weaponized

---

[273] This is why *voluntary* compliance and *voluntary* servitude is so important to "the powers that be."

version of a "science of the soul." With a basic understanding of the life unit and how it works as a bridge between 3D space and 3D time, one can unravel this process of cheating death, without having to resort to religious symbolism of Satan or Devil-worshiping.

Corporeal death is a "given," due to natural, involuntary or unnatural causes. Extending the life of the

*The Psychology of Souls*

body only has limited results and is not actually cheating death. What is actually done is to *engage the reincarnation process* to the point where the anima *disconnects* from the corpus (that is returning to inanimate status as chemicals), then drop that anima back in to *another body* that is young, alive and healthy. The problem to be overcome is that this young corpus is already "occupied." So, what is needed is either: an *uninhabited* corporeal body or a *tenant* that will voluntarily vacate the premises, when it is time for someone else to move in.

Our good friends of the New World Order have taken both approaches in their attempts to cheat death. First are the attempts to genetically engineer compatible bodies, from "alien hybrids" to modifications of life on our own world, just as the Annunaki originally did. Next comes the *volunteers*, people raised without any sense of identity or independence, so they do not have much, if any, of the spirit in the 3rd, ethical sector to put up any resistance when it comes time for eviction.

Bioenergy is still a form of energy, in general, and like its inanimate equivalent can be manipulated through *technology*—but requires an *organic* technology to do it. To the best of my knowledge, genetics has not yet advanced to the point of *creating* life, hence they are not able to engineer organic computers at the DNA level. So one must use the tools at hand, which is other life, combining their auras to act like a magnetic scoop, to catch a person that is dying and transfer that bioenergy into another corporeal body. These are "exposed" by journalists as those black, ceremonial rituals. It is *not* a lot of mumbo-jumbo, because it *is* generating vectors in 3D time—the cosmic sector where the anima remains intact as a structure. That is what the chanting and prayer is doing—aligning temporal vectors, like a net.

The consequence of these techniques is that the *same* souls with the *same* personalities, can continue on, indefinitely, just "downloading"[274] into a new body when theirs is about to expire or when it gets damaged beyond repair. Of course, the public does not suspect that we've got the *same* "world leaders" running things that we had 500 years ago, which explains why faces change, but the world continues to degenerate. They are cheating death and *avoiding karma* from reincarnation, so they do not have any concern over the consequences of their actions. As can be seen, depravity and criminal insanity tend to set in over the centuries.

### The Collective Soul

This ability to cheat death is partially genetic, in that it requires a stronger presence of Annunaki DNA to allow for a clean separation, since the SMs (our "gods") cheat death in a similar fashion—except their genetic engineering skills allow them

---

[274] In the re-envisioned version of *Battlestar Galactica*, the android Cylons were able to download their consciousness into a duplicate body upon death, making them immortal.

to grow their own, custom bodies[275] to relocate their souls into.

Whereas humanity is a hybrid of Annunaki and Neanderthal, this ability is reserved to the Blue Bloods, the Royal "children of the gods." The rest of the Red Bloods end up with a choice of reincarnation, with its karmic repercussions, or even ascension. Of course, because our world is controlled by those that *enjoy* fear and suffering, we all experience a great deal of *punishment*, as stated in the *Myth of Er*, and as such are continually *choosing to make a better life*—**we** are choosing to *evolve*.

What holds humanity back can best be described as a *collective soul*, the half that is tied to the Annunaki genetics that influences us to be voluntary servants and to not seek independence.

### Evolution of the Species

So where does this path of life, death, reincarnation, ascension, bioenergy, auras and psychology lead to? A doorway that is the next stage of human evolution, the *ethical man*,[276] taking his place as a peaceful explorer of the Universe. So why is this happening, now?

First, a quick review of "intervention" upon our world by extraterrestrial visitors, of which there were three major groups, the *Cyclopeans*, the *Titans* and the *Annunaki* (descendants of the Titan An/Anu), followed by some "locals" called the *Cabal* these days (the descendants of the Annunaki), along with the local inhabitants whom have been here all along, the LMs:[277]

---

[275] As depicted on many Egyptian etchings of the gods that are partly human, partly something else.

[276] In *The Tomorrow People* series (1973), this next stage of evolution was original called *homo superior* (superior man), then later changed to *homo novus* (new man). In the remake of *The Tomorrow People* (2013), they reverted back to *homo superior* so they could complain about it… after all, how could a bunch of kids be "superior" to the rivalrous, competitive politicians, bankers and priests running this planet? *Homo sapiens ethicus* is a more accurate term, reflecting the "ethical," Sector 3 origin of the evolution and meaning "the wise, ethical man."

[277] LMs refers to the "mythological" beings, such as faeries, dwarves, sprites, &c., along with the great apes, cetaceans and Neanderthal precursors; basically all the native life on the planet that evolved to obtain a higher degree of consciousness.

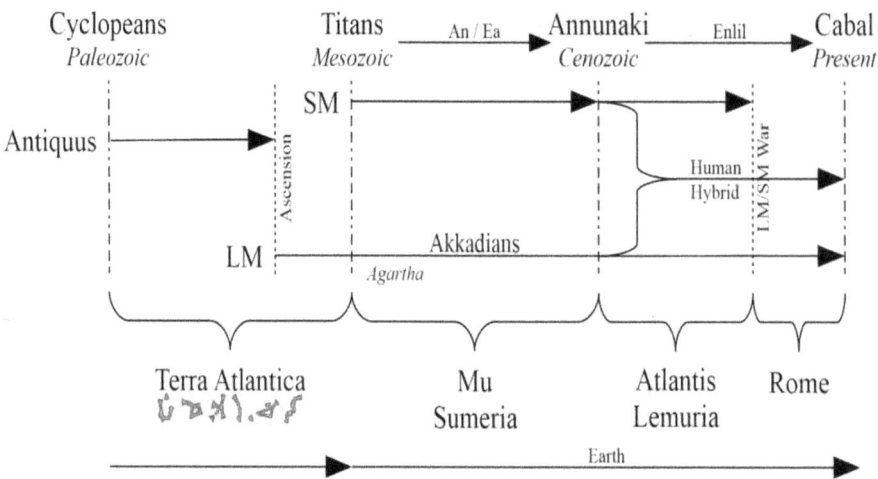

Tiamat

(Terra) (Ea's Eridu, "Home Away from Home")

As mentioned earlier, the *Cyclopeans* were an advanced, philosophical species that came here looking for type of monastic environment, somewhere "out of the way" to complete their research on ascension to the next octave of existence. In those days, our world was called *Tiamat*,[278] meaning "the final place of our love." The presence of these highly advanced beings, because of the nonlocal effects of bioenergy and auras, accelerated evolution on Tiamat causing the "explosion of life" that heralded the beginning of the Cambrian Period of the Paleozoic Era. Upon completion of their research, they left Tiamat by ascending to the Theta Octave,[279] leaving their monastic cities behind along with an archive[280] of their travels and researches—a kind of road map for those who may choose to follow. Long after the Cyclopeans ascended, another group arrived in our solar system— but not with such spiritual aspirations. They came here to exploit the natural resources of a newly developing solar system for more "militaristic" reasons, as our mythology speaks of: the war-mongering Gods of old, the *Titans*.

*Evolution of the Species*

Tiamat was not actually their destination; they just stopped by to make a survey, then proceeded on to Mars, which, being further out from our early, orange sun (remembering that astronomy is backwards), was further along the evolutionary path than Tiamat was, making it more suitable for colonization. And the Titans colonized Mars, using it as a base to access the enormous heavy metal resources of the asteroid belt.

---

[278] *ti+amat*, "our" + "love or fondness." *Ti* is also the final note of the diatonic scale, before the next octave begins.

[279] We currently exist in the *Eta Octave* of the seven densities, which is the 7th octave of the evolution of consciousness.

[280] This "Archive of the Ancients" is referred to as the "little red brick schoolhouse" in the research of George HuntWilliamson and can be accessed by any life that has developed a level 3, ethical consciousness. It also became popular on the *Stargate SG-1* television series as the repository of the Ancients (though taken off-world).

As the millennia passed, Mars had become fully established and, as Orson Welles so aptly put it, "*Yet across an immense ethereal gulf, minds that to our minds, as ours are to the beasts in the jungle, intellects vast, cool and unsympathetic, regarded this earth with envious eyes and slowly and surely drew their plans against us.*"[281] "Us," in this context, being the native, LM inhabitants that have been growing and evolving on Tiamat, courtesy of the Cyclopeans. The Titans dispatched An and his progeny to Tiamat to begin colonization under the direction of Enki, chief engineer and scientist. The colony was known as *Eridu*, in Mu (Sumeria). The planet became known as Ea's Eridu,[282] later truncated to *Earth*.

The colonization, exploitation and subjection of the "Earth" was the responsibility of An's progeny, the *Annunaki*. They founded the original colony in the motherland, Mu, then expanded over to the Americas/Antarctica as the Atlantean era, where they ran into some trouble—the barbarian rebels, with assistance of one of their own, Enki, who became technologically advanced enough to put up a considerable amount of resistance to the Annunaki "Gods." And these LMs did—and after a centurieslong, but successful attack on their "home world" of Mars,[283] forced the Annunaki off Earth.

Humanity has literally been "caught in the middle," as a hybrid of SM and LM genetics. The "pure line" of descendants from the Adamic stock became the overlords of the planet, the *Nobility, Royalty* or *Blue Bloods* that were the original children of the Gods, a mixture of Adamic and SM genetics. The remainder of the hybrids interbred with us low-life Neanderthals, becoming more like the Red Blooded natives, then slaves to serve the Gods. And that is where you find us, today. This "Cabal" of selfappointed rulers of the planet, keeping all the knowledge and secrets of their SM ancestry and technology to themselves, with the "peasants" to do their bidding.

Over the centuries, *homo sapiens*, the "thinking man," did exactly that. Man was genetically engineered with a significantly higher intelligence than the great apes and easily became the top of the food chain —namely, the best killers on the planet, courtesy of those SM genetics. This boosted intelligence was split over the two realms of biological life, half went to the *corpus*, embodied in the brain, what is termed *thinking* coupled to the physical *sensations*. The other half went to the anima, expressed in the heart—*feeling*, coupled to the temporal senses as *intuition*. These concepts form the basis of psychological typology as rational (thinking, feeling) and irrational (sensation, intuition) types.

Occasionally, mankind would think and feel at the same time, bringing balance to

---

[281] Welles, Orson, *The War of the Worlds*, Colombia Broadcasting System, October 30, 1938.

[282] There is evidence that "An" was a *title*, not a name, and the entity's name was *Ea*. That way, Ea's Eridu would always refer to the entity—not the job—in historical accounts, much like July is for Julius Caesar and August is for Augustus Caesar— we don't have a month called "Caesar."

[283] The war was basically an asteroid-throwing contest known as the *Great Bombardment*, leaving both planets covered with impact craters. In those early days, Mars was a fertile world, with oceans and an abundance of life.

both halves of the biological organism and realizing that these psychological valuing systems were just *two aspects* of something else, some kind of ethical or spiritual side manifest in physical reality, and the study of *Alchemy* was born—an attempt to document and reproduce this access to a kind of "spiritual motion," access to Larson's Sector 3 "control units" and the ethical level of existence.

Alchemical, spiritual knowledge[284] gave mankind the ability to advance beyond being a slave and automaton to the Nobility[285] and he started to do just that, with some covert assistance from the LMs, as far back as the 16th century.

This, however, was doomed to failure, as population levels were still small enough that the occasional plague could handle any insurrection and put the people back into a state of fear, running to the nobles for protection behind their large, impregnable castle walls.

In order to compensate for the developing spirituality in mankind, nobility was split into two factions: *governments* and *religions*—again, two aspects of the *same* nobility. So when people became upset with government, they could turn to religion to save them. When religion started the Inquisitions and the people became upset again, they could run back to government to save them. This oscillation has been going on for centuries; it takes about two centuries for people to get upset with one aspect, to switch over to the other—but still under control of the *same group* of nobles.[286]

In current world affairs, led by the *United States of America, Inc.*, we have the same situation, yet again. People are fed up with government lies, spying and abuse and are turning to the Christian religion to save them, demanding to "live free" under God (otherwise known as Enlil). There will be a short time of prosperity at the

---

[284] Alchemy derives from *Hermetics*—the god Hermes—one of the Annunaki that sided with Enki and the rebels that worked to assist mankind toward evolving into his own species. The Maya, Aztec and Inca, being the locals of Atlantis at the heart of the rebellion, all had a developed version of Hermetic knowledge that the Europeans brought back to the old world, where the actual tradition of Alchemy was born.

[285] Originally, there were four castes: the *nobility* (Royalty, being both political and religious leaders), the *military* to enforce the demands of the nobility, the *merchant* "delivery boys" and the *peasants*—the farmers and ranchers. Note that these form the four suits of the Tarot cards, "trumped" (or *triumphed*) over by the Annunaki.

[286] Wong, Eva, *Tales of the Dancing Dragon: Stories of the Tao*, Shambhala, 2007; an excellent historical account of Chinese history and this oscillation between government and religion; one comes into power with promises of greatness, becomes corrupt after a couple generations and is replaced by the other, doing the same thing. Contains many excellent stories of how seekers have reached this Level 3 consciousness and the attempts to control that information by both governments and established religions.

switchover, then back to the same, old thing. Just one dictator replacing another—unless mankind decides, "enough is enough," and makes a *conscious choice* to take the next step in human evolution: dump rivalry, embrace rapport and become peaceful explorers of the Universe in all its forms, as: *homo sapiens ethicus*, the ethical man.

**Humanity: The Next Generation—*Homo Sapiens Ethicus***

We have, right now, an excellent opportunity to make this transition to peaceful explorers, courtesy of a single, technological improvement that, curiously enough, was designed to make warfare more efficient: the *World-Wide Web* or *Internet*. For the first time since mankind was created, he has access to virtually *all the knowledge on the planet*, at the push of a button. The Internet has done to the Cabal, what the printing press did to the churches.

In the old days, the churches controlled what knowledge the general public had, through its scribes. The printing press allowed anyone to publish to the public, at little cost. The churches lost control of information and the public became smarter. The Internet is doing the same thing with the Cabal; no longer can they get away with false flags and corruption, without it showing up on YouTube. Only knowledge will set you free.

*Humanity: The Next Generation—Homo Sapiens Ethicus*

The historical path of evolution is summarized in the chart below, showing the paths of ancestry— nobility can trace its origins back to the Annunaki Gods, whereas the peasant stock only has an indirect route to the Gods (hence the necessity for an intercessor figure, like the Blessed Virgin Mary), with its roots planted firmly in the LM stock of ancient Tiamat.

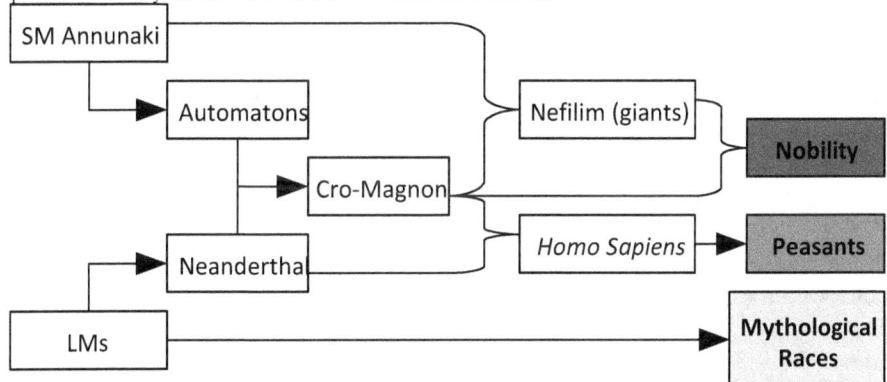

The life units that define the "noble body" are primarily from *genetic engineering*, as an engineered species can be custom designed for specific tasks and duties, from rulership to tripping over ottomans. Nobility *desires* genetically engineered life, as that is their path of evolution.

The "peasant body" tends to be sturdier, having a natural evolutionary path via "survival of the fittest" from the Neanderthal stock. The peasants desire natural processes and tend to be *organic* in their desires, as it is their path of evolution. Our planet also contains a mix of the two lines of descent, so there is a full range of expression of life.

So where is mankind going from here? Two choices are apparent, particularly if one looks at the goings-on in the world about them:

1.    A reinforcement of the Noble/Peasant system, by genetically engineering the peasant stock to be more "slave friendly" to the Nobility. Thank the "Gods" for Monsanto… literally!

2.    An evolution of the peasants into an *independent species*, a species that is basically organic in nature, does not need nor desire bioengineering. Like its LM ancestors, works in rapport with the other life of the planet (and elsewhere, as peaceful explorers).

Implementing of the first choice is simple enough, just "engineer" the food that limits the body and regulates the soul to encourage the system of rivalry and obedience, to get your good slaves. Also, reduce the population to manageable size, to go back to the good 'ole days of a single plague to keep the peasants from revolting.

The second choice is where many people are heading, and if you are reading this Paper, you are probably in this group seeking the evolution of the species.

So how does one do this? Simple: make the *ethical choice*! Work to develop your *shen* bioenergy and *become intelligent*. Forget the distractions that plague the "entertainment" industry and start learning new things. No one can do the work for you, so you will have to *choose* to become intelligent—the only thing that holds you back is societal programming, as it was designed to keep you "the Fool." Become the Magician, instead… learn, teach and advance yourself to the next stage of human evolution, that which Roger Damon Price referred to as *The Tomorrow People*, *homo sapiens ethicus*.

"Door of opportunity have sign that say, 'push to open.' It never locked."[287] *Alohomora*!

### *Epilog*

As some final comments, I want to point out that you are *not alone* on this *Magnum Opus*—the Great Work—to *homo sapiens ethicus*—there are many, many people now getting to this stage. And there is some assistance on this Path—not from more extraterrestrial aliens, but from right here on the planet— a little place that is the *anima* of the planet, called *Agartha*. It is said to hold all the beauty of Aphrodite. And if you've understood what I've said about life, the universe and everything, then I don't need to tell you where it is located, as you will have already figured it out for yourself—which is what you are *supposed to do*. To *do* the work, yourself.

The location of Agartha was chosen because it is *inaccessible* to the Titans, Annunaki and their Noble descendants, due to the structure of their soul. As such, it made for an excellent "rebel base" for the natives of Tiamat, and as things developed, transformed into a center of culture and philosophy for all the life on the planet—including the ethical men that have "figured it out" over the centuries. But, as explained in the earlier sections of this Paper, there is a policy of "non-interference" in terrestrial life, so they do not "channel" or send telepathic signals trying to influence people (or any of the other life on this world). Free will is paramount in that you are free to choose your Path, whatever path that may be. However, if you understand the *process*, then you'll know what "counts" as interference and non-interference.

---

[287] Singh Li Peng, Shaolin priest, China.

So learn what it is to be "human" and make use of the abilities that come with being human. Use your genetic heritage to advantage. There is a lot more to each and every person than they may realize—all that is needed is the opportunity for expression. Everything is based on *conscious choices*, so *choose to be wise* and seek the path that will improve life, not just for yourself, but for everyone you interact with. You will notice a change—and an opportunity.

# THE COLONIZATION OF TIAMAT

**A short time ago in a galaxy close, close nearby...**

*It is a period of scientific war.*
*Rebel researchers, working from a hidden basement, have won their first victory*
*against the evil World Order.*
*During the battle, rebel spies managed to steal*
*secret plans to the Order's*
*ultimate disinformation campaign,*
MODERN *ASTRONOMY, an armored system of theory with enough*
*assumptions to destroy an entire postulate.*
*Pursued by the Order's sinister agents, the ghost of Lloyd Pye*
*races home across the*
*cosmic sector, custodian of the insider info that can save humanity and restore*
*freedom to the galaxy...*

## Introduction

"Astronomy is backwards? Ridiculous!" Well, it seems the situation is a bit worse than even Dewey Larson realized, when he presented this "backwards" model of stellar and galactic evolution in his book, *The Universe of Motion*.[288] This was discussed, along with the planetary consequences thereof, in *Geochronology*,[289] the first part of this series. Larson's model was based on conventional astronomical data and the information presented in Part 1 was a *natural consequence* of that data. But what if that *data* is wrong? Well, not exactly "wrong," but viewed through a magnifying glass, making everything *appear* a lot bigger than it actually is, including the distances to other stars, the size of stars, themselves, and most importantly, the size of galaxies? What if stars *aren't* light years away, but light weeks away? "ET phone home" is no longer a long distance call, just a shout out the window. This Paper, Part 4 of the *Anthropology Series on the Hidden Origins of Homo Sapiens*, is

---

288 Larson, Dewey B., *The Universe of Motion*, North Pacific Publishers, 1984.

289 Phoenix III, Daniel, *Geochronology: Hiding History in the Past.*

an analysis of the extra-terrestrial influences that made this planet what it is today. The concept is known as *Intervention Theory* by the late Lloyd Pye,[290] speculating that our world has not only been visited by other species and civilizations, but it was actually *commonplace*—and they appear to have lacked Star Trek's "prime directive" of non-interference—they got their fingers in everything.

The primary objection to Intervention Theory was that habitable star systems were just too rare and distant for all these visitors to be coming to our world, known as *Tiamat*[291] in the ancient texts. But what if that was not the case—what if solar systems were *common* and *close by*, but just overlooked by astronomers and misinterpreted as something else?

An accidental discovery[292] was made by Reciprocal System researchers, while attempting to model the stellar neighborhood using Larson's concepts of *progression* and *scalar motion*. All the equations were checked, entered, the data loaded and the simulation was programmed to show the night sky—but when the results came out, the sky was black—not a single star, except for the points of lights representing the planets. Makes for a very lonely Universe.

No mistakes were found in the program code, but some debugging revealed the reason: all the stars were outside the *gravitational limit* of our sun. The Reciprocal System is based on discrete units of motion (quanta). When a net motion like the pull of gravity drops below *one* natural unit, it becomes *zero* and disappears—there are no fractional parts.[293] In the RS, the reach of gravity is limited to that distance—and no further. Gravity has *no effect* beyond this "gravitational limit."

Another consequence, documented by both Larson and Prof. KVK Nehru of India, is that a 3-dimensional coordinate system can only exist *within* the gravitational limit. Once you go past it, the *progression*[294] takes over and the loss of dimension reduces "space" to "equivalent space," a 2-dimensional, $c^2$ form of space that is analogous to "hyperspace" in Science Fiction. What made the stars disappear in the simulation was that the progression is a *scalar expansion* at the speed of light— and it got that label because the photons were being *carried* by it, so when we measure the speed of light we are actually measuring the speed of the outward progression—the photons *do not move* relative to that progression. The light from the other stars could not cross the progressive void that existed

---

290 Pye, Lloyd A. Jr., died of lymphoma cancer on December 9, 2013.

291 *Tiamat*, translation: the place of our love.

292 See: "Visibility of Stars and Planets (Problem)" topic in the RS2 forum: http://fora.rs2theory.org

293 In conventional astrophysics, there is no limit to the pull of gravity—it just gets smaller and smaller, all the way out to infinity.

294 The "progression" is a scalar expansion of space that is actually recognized by conventional astronomy as the "Hubble Expansion."

between the gravitational limits of the stars and the simulation determined that stars were out of visible range and did not render them—resulting in the black sky. The planets, being inside the gravitational limit, *did* render as points of light, as did the sun.

Attempts to understand and solve this problem led to the research of the late Behram Katirai's[295] book, *Revolution in Astronomy*. Katirai, along with many amateur astronomers, wondered as to how we can see the light from objects so incredibly distant. To claim that the Andromeda galaxy can be seen with the unaided eye, some 2.5 million light years away, is counter-intuitive. Plain "common sense" says there is something wrong with this picture. We can barely see our own Milky Way galaxy through the light pollution of the sky—can you imagine how bright Andromeda must be, to be able to see it at that distance?

*Introduction*

Katirai determined *exactly* how far a human being could see, with or without the aid of telescopes. He focused on the Hubble Space Telescope, concluding that its maximum range was a meager 357.14 light years. Recent upgrades, including digital imaging, may have increased that distance 10-fold, but even 3571.4 light years is still far short of *ever* being able to see galaxies that are millions or billions of light years away—that is, unless those galaxies are a LOT closer than we think they are.

That is the conclusion that both Katirai and the recent Reciprocal System research into the gravitational limit of stars has concluded: these "galaxies far, far away" are actually "*solar systems* close, close nearby." And for those of us that remain "uncommitted investigators" and are able to *actually consider* this radically new concept—and the estimated *100 billion* "galaxies," a.k.a. "solar systems" within range of our existing, "3571 light year" telescopes… the implications are staggering.

Larson is not infallible and took a "short cut" in his astronomical research, using the data provided by the astronomers of the 1950s to base is stellar calculations upon, rather than attempting to derive stellar and galactic geometry directly from his physics of motion.[296] Logically, his system works as explained. However, because he leaped from atomic data to stellar data in a single bound, he skipped over some important, intermediary consequences that solve the puzzle of what it is that we actually see in the night sky. Larson's entire theory is based on the concept of *scalar motion*, and funnily enough, his mistake was one of *scale*!

The original problem was this: the Universe is constantly expanding. Gravitation

---

295 Katirai, Behram, 1948-2010.

296 Larson did use his atomic physics to determine that stellar evolution, and therefore galactic evolution, was backwards because of the nature of the energy generation of stars. His redshift work on quasars and pulsars was also derived from atomic physics, since quasars were not even discovered until the 1960s, some 10 years later.

is the inverse of that expansion—compression.[297] This "Hubble Expansion"[298] wants to push everything apart at the speed of light, whereas gravitation wants to pull everything together at the speed of light. In a gravitationallybound system, there is a balance between the outward expansion and the inward compression, giving us our conventional, 3D reference system. Beyond the gravitational limit, the outward expansion wins, hands down. Consider the case of an expanding balloon with spots drawn on it. As the balloon expands, all the spots get further apart from all the others—this is *scalar* motion, a *change in scale*. If you add a spot in between, it still moves away from all the other spots and will never run into one. This is the case with photons and the "space" outside the gravitational limit. They are just new photon "spots" and are moving away from each other *and all the stars*—light cannot cross that void.

The way to fix the problem was to eliminate the expansion zones between stars, which meant reducing the observed distances between stars—considerably. This "begged the question" of why these distances, originally measured through triangulation, were off by so much. The triangulation method was simple enough, measure the angle between the Sun and the star to be measured. Wait 6 months until we were on the opposite side of the sun and take another measurement so you had a triangle with a known distance at the base (2 AU) and two angles to the sides. Later on, "stellar parallax" was used, which is a similar concept that measures change relative to background stars (that are assumed to be stationary and that those stars are actually *behind* the star you are trying to measure). This is how we obtained all our stellar and galactic distances.

But the gravitational limit, a concept unknown to modern astronomy, was not taken into account. Space was assumed to be 3D and homogeneous everywhere, which is actually not the case. Studies into globular clusters[299] and galaxies reveal that the system moves like a viscous liquid or hot solid—not objects floating around like gas molecules in a void. All the stars and nebula appear to be bound together, rather tightly. Galaxies appear to have much the same structure and behavior as the resulting whirlpool in your sink, when you pull the plug on the drain.[300]

---

297 Gravitation is "inward in space" and because of the reciprocal relation, also "outward in *time*." In the RS, all atomic rotation is "outward in time," and therefore produces an "inward in space" gravitation, directly from its structure.

298 KVK Nehru, "The Gravitational Limit and Hubble's Law," *Reciprocity* 16 № 2 page 11.

299 KVK Nehru, "The Large-Scale Structure of the Physical Universe: The Cosmic Bubbles," *Reciprocity* 20 № 2 page 5.

300 According to a story by Dewey Larson's daughter, Linda, her father came to an understanding of galactic rotation by watching his shaving cream and

This observation provides an opportunity to understand the effect of the gravitational limit on what we see. In essence, the 3D region inside the gravitational limit has a structure similar to a clear liquid, whereas the region outside the limit is more like a gas. We now have a simple analogy to understand what is going on: jump in a pond and look up at objects in the air. If you've ever been diving or swimming off a boat, you may be familiar with the problem of the *index of refraction.*

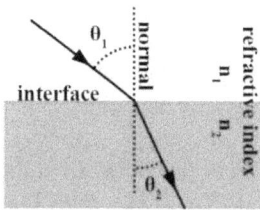

If you approach a boat from underwater, reach out with your hand and try to grab a rung on the ladder on the side of the boat—you miss, because the rung isn't where you see it. When light moves between mediums of different density, it bends—refracts—but your arm does not. The same problem occurs when trying to catch fish with your hands, standing in water. You clearly see the fish, grab for it and miss—because the image of the fish isn't where the fish actually is.

This is exactly what is happening between the 3D "water" space inside the gravitational limit, and the 2D "air" space outside the limit. The stars we see in space aren't where we think they are—the actual object is much closer than we observe it to be, which has led to these miscalculations of distance. Index of Refraction at Gravitational Limit ignored

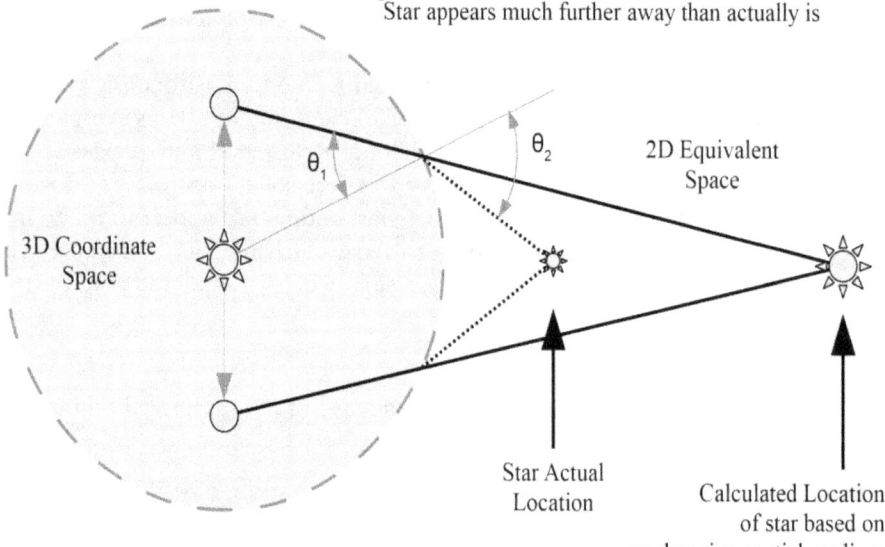

**Astronomy 2.0**

whiskers spin down into the drain of his bathroom sink, realizing that galactic cores were *consuming* stars—not creating them. (Creation was from globular clusters.)

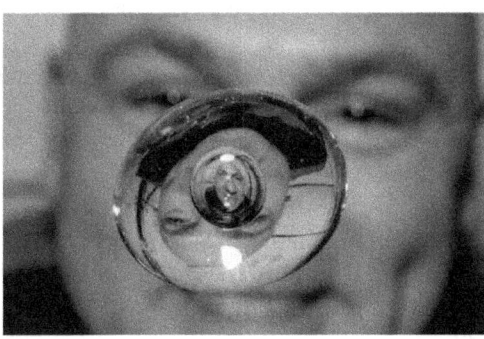

So where does that leave astronomy, if what we've been told is wrong? In essence, astronomers have been looking at the Universe through a fish-eye lens that has been magnifying things out of all proportion. Add that to the backwards evolutionary processes and, well, you can see the trouble this science is in.

But here is what we *do* know: the objects within our solar system have been measured to a reasonable degree of accuracy. In the Reciprocal System, everything works the same way, so what we see in our solar system is probably typical of other solar systems, as well as both larger and smaller constructs. By that, I mean if you look at our solar system, you find a collection of planets in a roughly planar orbit, asteroid and Kuipner belts, and at a far distance, the Oort cloud forming a sphere of debris around the system. Now compare that to Jupiter—a miniature version of a solar system. Again, you have moons in a roughly planar orbit, ring systems that are "belts," and at a far distance, small asteroids and moonlets orbiting around the planet in a rougly spherical distribution. The Jupiter system is just our solar system in miniature—a scaled-down version.

And this is the situation we find with the observed "stars" and "galaxies." When we scale them down, we find that stars are actually Jupiter-class gas giant *planets*, and "galaxies" are solar systems. As revealed by Hubble photographs, the cores of galaxies appear as *single stars* (sometimes doubles or multiples, mostly single). The encircling cloud of the Milky Way galaxy is just the outer rim of OUR solar system—not an independent object. This is also evidenced by the difficult time astronomers have had, trying to *find* the core of the Milky Way galaxy. Eventually, they decided it was in Sagittarius, Sgr-A, because that spot was lit up a bit more than the rest.

Take a look at this artists conception of our solar system (left), then compare to an actual photo of the Sombrero Galaxy (right):

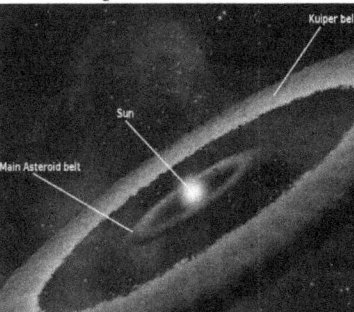
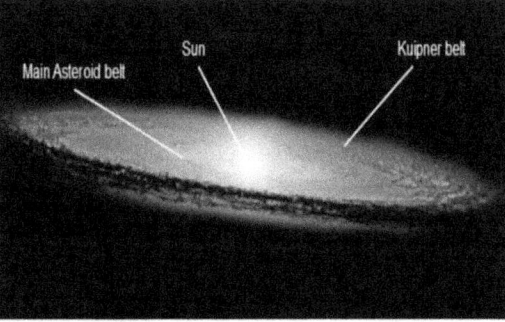

If you have never seen a picture of the Sombrero galaxy before and some came along and showed you this photo, stating that "the Hubble found a new solar system!" ... you'd probably believe it without question, because it *looks* exactly like a solar system—not a galaxy. Bring up your favorite search engine and take a look at the Hubble Space Telescope photographs of galaxies, in all their remarkable detail, then ask yourself, "am I looking at a galaxy, or a newly forming solar system?" Check the details; the "galactic core" looks like a single sun, complete with a defined disk. Asteroid belts are commonplace, along with a lot of dust and debris—what Larson says provides the fuel to power the sun, through atomic fission. (Our sun is older than most of these others, hence it has digested most of the debris that would be found between the planets. Of course, the planets also accumulate dust and debris. The Earth, alone, sucks in about 100,000 metric tons of cosmic dust every year—and that is low because most of that dust and debris has already been captured.

Observation has come a long way since the early, black and white photographs. As resolution improves, what was once thought to be light from millions of stars is now seen as luminous gas clouds; nebula become sharp and defined and the large blob that was once the galactic core has significantly reduced in size, to a small sphere. As resolution improves, the core will probably become smaller and more defined, until it looks exactly like the sun it may well be.

Andromeda (circa 1950)  Andromeda (Hubble)

These natural consequences create a very different picture of the Universe in which we live. We are no longer isolated and alone, but in the midst of many thousands of solar systems that all have similar conditions to our own—life may not only exist "out there," but in all probability, will be very abundant.

This simpler view of astronomy solves a number of problems with both conventional theory and Intervention Theory:

1.     There are *thousands*, if not *millions*, of "galactic" solar systems within range of our telescopes.[301]

2.     They are nearby and within range of Ark[302] technology, as well

---

301 The Hubble Space Telescope has photographed over 3000 of these solar system "galaxies" in just a small section of space. Extrapolations run from millions to 500 billion "galaxies" in observational range.

302 Arks are spacecraft constructed from supernova asteroids, where the ultra-high speed motion creates a hollow core, similar to a Dyson sphere on a smaller scale, that acts as a self-contained habitat for long space journeys. Our moon and the Martian Phobos and Deimos are all Arks. The Lunar Ark

as our own, electrogravitic spacecraft. (Refraction-corrected approximations put the Andromeda "solar system" a mere three light *months* away.)

3.      "Supermassive black holes" are no longer needed to hold together billions of stars in a "galaxy," since it's actually just a small, solar system like our own.

4.      Wormholes and warp drive become unnecessary; travel between stars takes about the same amount of time the ancient mariners took to sail between Europe and America.

5.      No longer a need for "dark matter," to account for missing mass, because the scale was wrong and there is no mass missing.

6.      No need for "dark energy" to push things apart; the progression of the natural reference system already accounts for that, in totality.

7.      Many points of light we see in the sky are just *asteroids* being lit up by the sun, typically highly reflective gas giants formed from the supernova debris field that created the solar system.

*Astronomy 2.0*

8.      What we now view as the *Milky Way* **galaxy** is just another solar system that is gravitationally bound to our own. Many "galaxies" are in the same situation; see NGC 4674 A&B, NGC 7318 A&B, all the "Antennae" and "Mice" galaxies. They are just solar systems within the gravitational limits of each other.

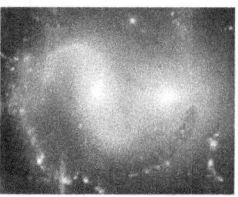

9.      Other "actual" stars have been misidentified as *quasars*. Larson's research into quasi-stellar objects showed that the redshift measurements were fudged by astronomers and the redshifts indicated these objects are actually quite close—not at the extreme limits of the Universe.

10.      Real galaxies do exist, having been identified as *quasar galaxies* (galaxies made of quasars). When scaled down, the quasars become the stars of the galaxy.

11.      The extreme orbital speeds of exoplanets become scaled down to moons orbiting Jupiter-like planets at the normal speeds observed in our own solar system.

As an example to item #11, we can take a conventional star with known exoplanets, such as Kepler101, a single sun with two planets, 101-b and 101-c. 101-b orbits this star in 3.49 days, and 101-c in just 6.03 days. The fastest planet we have in our solar system is Mercury, taking 88 days. That's a big difference. But what if we scale the star Kepler-101 down to a Jupiter-size planet? Jupiter has a bunch of moons and if it is a Jupiter-size planet, 101-b and 101-c should show *similar orbital properties* as some of Jupiter's moons.

---

Nibiru was used to transport the Annunaki here.

Jupiter is roughly 1/10$^{th}$ the size of the sun, so we can just adjust the orbital distance by a factor of 10:

101-b: 0.045 AU / 10 = 0.0045 AU.

101-c: 0.0648 AU / 10 = 0.00648 AU.

So, we are looking for a couple of moons at these distances, with similar orbital periods (the period is not scaled as it is time, not spatial distance):

**101-b**: 0.0045 AU, 3.49 days.

**101-c**: 0.00648 AU, 6.03 days.

Lo and behold…

**Europa**: 0.0045 AU, 3.55 days. Almost an *exact* match to Kepler 101-b.

**Ganymede**: 0.00716 AU, 7.15 days. Just a little further out than Kepler 101-c.

To paraphrase Obi-Wan Kenobi, "that's no planet, it's a moon!"[303] And a moon orbits a *larger planet*, not a star. What they found are *not* exoplanets around stars, far, far away, but *moons* around large *planets*, close, close nearby.

As hard as it may be to believe, it looks like the "experts" did it again and got everything wrong. I know how surprised you must be.

But with this new perspective of Astronomy 2.0, we actually live in a rather crowded section of space. Initial estimates place over 1200 solar systems within a "5 year mission" of electrogravitic ships (ships than can travel at near light speed). All these stories of UFOs and ETs may well be true, and with the potential of hundreds of different civilizations paying us a visit, it certainly explains the wide variety of different craft designs.

However, I still have my doubts about "channelers," because I could not help but notice that these hyper-intelligent, pan-dimensional beings never bothered to tell us that we've got our "scale" wrong and that galaxies were actually solar systems. Though it did appear that Ra, channeled through Carla Rueckert and questioned by Don Elkins in the 1980s, made the *attempt*, but we were so programmed into the galaxy-mode thinking that we never connected the dots… *The Ra Material* 16.35:

**Ra**: I am Ra. I see the confusion. We have difficulty with your language. The "galaxy" term must be split. We call "galaxy" that vibrational complex that is local. Thus, *your sun is what we would call the center of a galaxy*. We see you have another meaning for this term.

**Elkins**: Would you define the word "galaxy" as you just used it?

**Ra**: I am Ra. We use that term in this sense as you would use *star systems*.

One also has to wonder why some of the more open-minded astronomers have not noticed that there is something wrong with the data concerning quasars, pulsars, galactic motion and long-period comets, since the values must appear a bit ridiculous. Larson noticed the quasar redshift problem some 50 years ago, from published redshift data. When checking up on his redshift sources, I found that the High Priests of Astronomy have *changed the data*, to fit their theories! All

---

303 Star Wars, "That's no moon, it's a space station!" upon approach to the Death Star. However, this quote does apply to many of the moons in our solar system, which are "dead Arks" from other civilizations.

the early redshift data of assumedto-be remote objects had many of them receding at several times the speed of light (Larson documents recession velocities of up to six times the speed of light—impossible, in Einstein's universe.) This has now all been "adjusted" to hide that evidence.

So I looked around for other anomalies that may have been published by deep-sky researchers, particularly those involved with measuring galaxies and long-period comets (that may well cross between gravitationally-bound solar systems). Ran into a problem, however…

- Marc Aaronson, researcher into age and size of Universe, killed by revolving dome of telescope.
- Carol Ambruster, stellar researcher, murdered.
- David Burstein, expanding universe researcher, dead from Pick's disease.
- Richard Crowe, co-founder of Hilo's astronomy program, killed in car accident.
- Thomas Gold, researcher into beginnings of life on Earth, dead from heart failure.
- Robert Harrington, Planet X researcher, dead from cancer.
- John Huchra, published papers on a different birth of galaxies, dead from heart attack.
- Brian Marsden, comet and asteroid tracker, dead from cancer.
- Koh-Ichiro Morita, ALMA researcher, murdered.
- Steven Rawlings, assisting to construct the Square Kilometer Array, murdered.
- Allan Sandage, attempting to measure the rate of expansion of the Universe, dead from cancer.
- Eugene Shoemaker, cometary researcher, killed in car crash.
- Walter Steiger, site manager of the Submillimeter Observatory, killed in car crash.

Let's just say it is a somewhat extensive list and few people realize that looking through a telescope is a very high risk job! Speculation went around prior to 2012 that these people had stumbled upon Nibiru (Planet X) and were silenced, but since X was a no-show and the ancient, historical records indicate that the Annunaki arrived on *our moon*—that would indicate our own moon *is* the Ark Nibiru. I suspect what these late astronomers found was *bigger* than that—that astronomy, as we are force-fed it, is *totally wrong* and we are far from alone. But as long as we *feel* isolated and cut off from the Universe, we are a lot easier to control. *4.6 Billion Years Ago… I think it was a Tuesday…*

## 4.6 Billion Years Ago… I think it was a Tuesday…

As discussed in *Geochronology*, stars and planets are no where near as old as astronomers say they are. The calculations of Prof. KVK Nehru state, "… indicates that a star of, say, *one solar mass* would condense in $0.138 \times 10^8$ years"[304]

---

304 KVK, Nehru, "The Large Scale Structure of the Physical Universe,"

(13.8 million years). Our sun is "one solar mass" and would therefore be only about 14 million years old—not 4.6 billion. Correcting for the backwards stellar evolution puts the sun near the start of its life, not the end, having an estimated life span of about 80 million years.

When we look at the geologic and anthropologic history of our world, the time scales are correspondingly exaggerated. This exaggeration prevents "arm chair" researchers from noticing the correspondences between things, effectively *hiding history in the past*. This was the case with the CroMagnon man. When the 50,000 BC time line of their fossil record was updated to about 6,000 BC, all of a sudden, Cro-Magnon man appeared at exactly the same time as the Biblical Adam and Eve— whom were creations of the Annunaki—explaining *why* Cro-Magnon seems to have appeared out of nowhere.

Geologic history is broken down into supereons, eons, eras, periods, epochs and ages. The breaks between them are usually indicative of some kind of geological

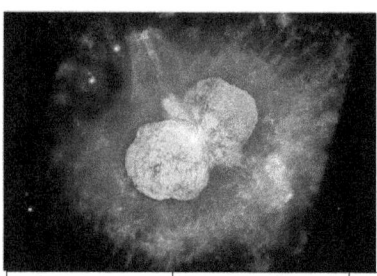

event, such as a solar change or planetary expansion event. The dates are wrong, but the groupings of who and what was there tends to be correct. There are four eons defined to section-out the 4.6 billion years of history (let's round up to 15 million, for some leeway in the correction). They are (times in millions of years ago):

| Eon | Conventional | Event | *Updated* |
|---|---|---|---|
| **Hadean** | 4540 | Post-supernova, primordial state of planetary formation. Sun is basically a ball of debris, so system looks like a nebula. | 15 |
| **Archean** | 4000 | Volcanic, Class M red giant sun, little atmosphere, no oxygen, basic prokaryota life. | 13 |
| **Proterozoic** | 2500 | Crust formation and start of lakes; eukaryotes generating oxygen, but highly radioactive surface. Many expansion events; Class K sun. | 8 |
| **Phanerozoic** | 542 | Class G sun; habitable surface. Arrival of the Cyclopeans. | 2 |

The first three Eons are not very significant from the perspective of life. The supernova explosion created an enormous debris field that the newly forming planets (from the destruction of the binary companion) would be plowing through for some time.

Gravity eventually condenses the debris field back into a class M red giant then continues to compress it into the orange, Class K sun.

---

Section 9.2 "Globular Clusters," Equation 21.

Because of the significant quantity of debris, this process moves fairly
quickly and before long, the new, orange sun has condensed and cooled
sufficiently, putting out about 75% of the energy that it does today, making the
surface of the inner planets, namely Venus, Tiamat and Mars, suitable for life—
and colonization.

But it is the Phanerozoic Eon where things get interesting.

**Arrival of the Cyclopeans**

Based on our updated geochronology, virtually all historical data obtained from
fossil records has occurred within the last two million years. The start of this Eon
is referred to as the *Cambrian explosion of life*, because from out of nowhere the
planet suddenly burst full of diverse life. Reminds me a bit of how Cro-Magnon
man showed up out of nowhere, just after the Annunaki arrival. But it this case it
was another race of giants, referred to by George Hunt Williamson as the
*Cyclopeans*.

The Cyclopeans[305] were a very advanced race of peaceful, interstellar explorers,
having a reputation of being excellent builders and craftsmen. They had explored
thousands of solar systems (galaxies) and, like the Ancients of the *Stargate*
franchise, reached a point in their evolution that put them on the point of
*ascension*—not to the next *density*, but to the next *octave* of existence.[306]

Though the Cyclopeans were not plagued with the constant political trouble-
making of the Annunakidescended nobility of humanity, they also found a need
and desire to establish their version of a *monastery*, a place of sanctuary and refuge
to focus on their research and personal evolution.[307] And that is what brought
them to our newly-forming solar system, some two million years ago.

---

305 From *cyclopes*, meaning one-eyed. Williamson states in his books that
the Cyclopeans did have a single eye, but it is difficult to believe that such an
advanced race would not have depth perception from the lack of
stereoscopic vision. Of the surviving pictographs, the Cyclopeans are
depicted as having one, large eye, which upon examination, could have
easily been the visor of a protective helmet, much like our own pilots and
astronauts use. The planetary environment was rather unstable in those
days, so protective clothing is definitely a possibility.

306 In esoteric philosophy, consciousness evolves through a series of
"densities" (levels of complexity) through the ascension process. Eight of
these densities form an octave, much like their musical counterpart. We are
currently in the Eta Octave of existence (the 7th). The Cyclopeans were
attempting to ascend to the first density of the Theta Octave.

307 A similar, human effort is being made by the Antiquatis Institute's
*Sanctuary Project* and *Kheb Monastery* (named after the Kheb reference in
the *Stargate SG-1* episode, "Maternal Instinct"–the place where the
Ancients left documentation on how to ascend for others that wished to
follow the same path.)

A simple way to understand what went on during this period is to look at the early, European settlers in American history. They did not arrive in the new world with supercomputers, Internet and cell phones, but with bags of grain, chickens, cows and horses—most of which were *not native* to the new land. The Cyclopean colonists did the same thing, arriving in our solar system, selecting this "3rd rock from the sun" as the site to build this sanctuary and started unloading their bags of grains—but from orbit, seeding the entire surface of the planet with life suitable for their needs. Finding fertile, volcanic soil and plenty of moisture about, life exploded across the planet, covering the land and the seas, creating "farmland" for the Cyclopean versions of chickens and cows, just like the New England farmers did after arriving from the Old World. And they named this new world, formed from the chaos of a supernova remnant, *Tiamat*, the "place of our love."[308] Every planet has its own, unique organisms and genetic paths, and Tiamat was no different. When settlers arrived in America there was already stuff growing there. After introducing the seeds and animals they brought with them, hybridization occurred, much like the combination of a horse and donkey producing a mule. And that happened on Tiamat with the Cyclopean stock—an explosion of diverse life, from the microscopic all the way up to many of the creatures we find in the fossil record.

With plenty of food, beasts and building material now available on the planet, the Cyclopeans started building their sanctuary. They were giants in stature (not quivering, purple tubes), standing some 5-7 meters (15-20 feet) in height, so their constructions were *megalithic* by our modern standards. Their

*Arrival of the Cyclopeans*

technology, based on what we term *vibratory physics*, allowed them to manipulate inanimate structures easily, such as dissolving rock or fusing it into glass much in the way that John W. Keely demonstrated with his "vibratory sympathy" machines of the 19th century. They constructed their facilities inside of mountains, opening passageways and fusing the sides into black, obsidian glass (a hallmark of their handiwork), connecting the facilities together with high-speed, underground transit systems. Upon completion of these world-wide facilities, they were able to settle in and begin their research on ascension to the Theta Octave.

---

308 Names tend to be reused, over and over, as is the case with Tiamat and the later applications from the Sumerian and Greek mythology. A Ford "galaxy" has nothing to do with what's up in space, though they have the same label.

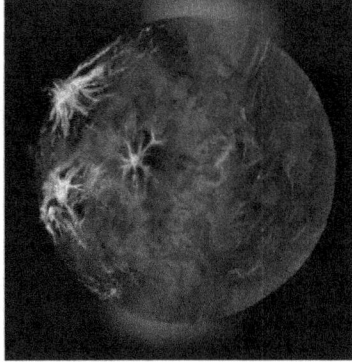

In Cyclopean days, Tiamat was physically about a fourth the size that the Earth is now, as it had yet to undergo any major expansion events. Also, being in a gravitational lock with the new sun that was getting hotter by the year, the surface was split into three regions:

- *Múspellsheimr*, the hot, volcanically active region facing the sun.
- *Niflheimr*, the cold, dark side facing away from the sun.
- *Ginnungagap*,[309] the habitable zone where fire met ice, having reasonable temperatures, lakes and rivers.

If you are familiar with Old Norse mythology, then you will realize that there *was* a race of ancient jötnar (giants) that spawned from (colonized) Ginnungagap, the *hrímþursar* (*rime thurs*, the *Frost Giants*).[310] Legends of this first race, some of whom were still around when the Titans arrived to plunder the resources of Tiamat, are found in just about every mythology on the planet with different names, usually translating as a reference to "wise old ones," such as the Elders, Elder Race or Antiquus.[311]

These Cyclopean jötnar were a dedicated group that understood the ways of Nature and natural processes, which was the point of their study of the Theta Octave and the path that led to ascension to it. They worked in harmony with the planet and as a consequence of that "harmonic resonance," life flourished on Tiamat, soon producing intelligent life. (Unlike the Annunaki genetic engineering, it was just the Cyclopean *presence* that caused consciousness to evolve at an accelerated rate.)

As Larson mentions in *Beyond Space and Time*, "anywhere life can exist, life does exist." And intelligent life sprang up all around the globe, on the land, in the sea, under the ground and in the air, remembered today as the "mythological

---

309 *Ginnungagap*, Old Norse, meaning "magical and creative, power-filled space." Snorri Sturluson's *Gylfaginning* states, "Just as from Niflheim there arose coldness and all things grim, so what was facing close to Muspell was hot and bright, but Ginnungagap was as mild as a windless sky."

310 The hrímþursar were later given a very bad reputation by the Æsir (the Vedic *Asura* or the Sumerian *Annunaki*), whom spread rumors about how they killed the entire race—which was their way to explain why the race *disappeared*— ascended, not killed. Based in rivalry, the Æsir prefer a victory over a defeat.

311 Ymir, the progenitor of the hrímþursar, is known in the Greek as *Uranus*. Uranus may be the origin of the word *Urantia* which has been also used as the name of the planet in early times.

creatures" of yesteryear. They evolved concurrently with the Cyclopean researchers, picking up many of the same habits; predominately the desire to work together as peaceful explorers of their world, living in harmony with nature and yin-based technology to *improve* the performance of nature (for example, raising water tables to irrigate crops through forest management, rather than pumping water out of the ground and distributing it in ditches).

Over time, the natural processes of the inner core of the planet and the sun caused many expansion events, cracking the single, rocky mass into continents and expanding the rivers into lakes, the lakes into seas and the seas into oceans. Life continued to adapt to the new conditions, evolving and becoming more diverse. The field of consciousness generated by the Cyclopean researchers tended to mold the more advanced life into a similar, bipedal form, giving rise to a large variety of species that had the *image* of the Cyclopeans, but in miniature: the races known to the "insider community" as the L-Ms, the "Little Men" (the mythological races of sprites, nymphs, faeries, dwarves and their kin).

It is not known if the L-Ms were just a natural evolution of Tiamat, if they spawned from the life brought to the planet by the Cyclopeans or if they were a hybrid of the two. What is known is that the L-Ms *evolved* from the lower forms on the planet. This has an interesting consequence that is unfamiliar to humanity—the L-Ms developed *spiritually* from the research drive of the Cyclopeans, but lacking a "creator god," never developed the concept of *religion*. Not being *engineered* as a slave race, they do not *worship* anything, nor anyone. Their spirituality is based on the *evolution of consciousness*, which was how they came into being, just as human spirituality is based upon worship and servitude, which is how humanity came into being. To quote Mr. Spock, "fascinating."

The Cyclopeans, being the master builders of the Universe (I suspect Freemasonry has some of their concepts from the Cyclopeans, distorted by rivalry), took advantage of the expanding planet and the new territory being made available within the depths of its interior. The structure of planets is much like that of the Arks and white dwarf stars, having a hard, crustal shell then a diminishing density gradient with a hollow interior. But in the case of planets, which are living organisms, that hollow center exists past the ultra-high speed range described by Larson—it is cosmic, existing in 3D time, appearing inside-out to 3D space dwellers. If one were to enter the hollow core, they would actually find themselves on the surface of the cosmic aspect of the planet, it's "soul." And that's what the Cyclopeans did, since their vibrational technology allowed for easy and compatible access to this realm, being assisted by the L-Ms, whom were always interested in discovering new things.

It was within this hollow core, the surface of a cosmic world, that the Cyclopeans constructed their monastery and archive; a record of their journey across the cosmos and the path to ascension out of the octave, as a legacy for the L-Ms (and other life that may develop). This monastery, surviving in mythological records as *Agartha*, protected by the Elementals and underground dwelling L-Ms, became the Cyclopean's final departing point for the Theta Octave.

They were in no rush to ascend; there was still a lot to learn in this Octave and, for them, it was about learning, growth and the evolution of consciousness. The passage of time did not matter to the Cyclopeans, as they had already evolved past the need for the cycle of reincarnation (except in case of accidental death) and were technically immortal. So with cities in the mountains on the surface and a repository of knowledge and sanctuary deep in the hollow core of the planet, they had completed this research outpost and could get down to serious work, for them, and their kin, many of whom were still scattered about the galaxy.

But a mere million years into their research project, strange lights were seen in the sky—visitors from another solar system, close, close nearby... the *Titans* had arrived.

# THE COLONIZATION OF TIAMAT:
## *THE ANNUNAKI STRIKE BACK*

*It is a dark time for the*
*Rebellion. Although the misconceptions of modern astronomy have been destroyed,*
*conventional astronomers have driven the*
*Rebel forces from their hidden basement and pursued them across*
*the solar systems, still claiming they are galaxies.*
*Evading the dreaded Royal Astronomical*
*Society, a group of free thinkers led by Daniel Earthwalker has established a new secret*
*basement in the remote ice state of Montana.*
*The evil lord Darth Enlil, obsessed with finding young*
*Earthwalker, has dispatched thousands of seraphim into the far reaches of the wilderness…*

Many common sense concepts accepted as *truth* in the 19th century have now been delegated to the realm of "poopoo" by our hyper-educated, modern scientific thinkers. And so it was with the concept of the *æther*—or more accurately, being so embarrassed to discover that these 19th century researchers were actually *right*, these scientists hid the concept in the dark—*dark matter*,[312] to be specific. None of the 19th century æther researchers are mentioned in dark matter research so these experts can get their own claim to fame for this brilliant, new, centuries-old idea to explain why the Universe isn't behaving the way it should… according to modern theory.

To assist in the later developments of the colonization of Tiamat, two concepts are going to be introduced, again using the context of Dewey B. Larson's *Reciprocal System* of theory:[313]

    1.       A new concept of *æther*, being that of the projection of Larson's

---

[312] Wikipedia: "Dark matter is a type of matter hypothesized in astronomy and cosmology to account for effects that appear to be the result of mass where no such mass can be seen. Dark matter cannot be seen directly with telescopes; evidently it neither emits nor absorbs light or other electromagnetic radiation at any significant level."
Which happens to be the exact description of æther in 19th century texts. Dark *energy* is normally associated with æther, but that is simply Larson's *progression of the natural reference system*.

[313] Books and papers on the *Reciprocal System of theory* can be downloaded for free at http://reciprocalsystem.org

*cosmic sector* (the realm of *3D, coordinate time* and *clock space*) into *yin space* ("equivalent" or vortex space).

2.        The *hollow planet theory*—not just a hollow Earth, they are *all* that way—stars and moons, too!

But as usual, conventional thought tends to be a bit backwards—or in this case, inside-out.

### Æther Theory

*"There is no space without aether and no aether which does not occupy space."*
*– Sir Arthur Stanley Eddington, astrophysicist, 1882-1944*

The Reciprocal System is based on a reciprocal relation between space (s) and time (t) that we conventionally refer to as the ratio of *speed* (s/t) or *energy* (t/s). In the universe of motion, that is *all* you have to work with—space and time—*nothing else*. With only two choices it greatly simplifies understanding, because we already know all about 3D space with clock time. We observe space as a *vacuum*, with *stuff* (particles, atoms and molecules) in it. Since space is *empty*, that "atomic stuff" *cannot be space*—so the only other choice is *time*.

Larson's atoms are simply a *temporal rotation* in three dimensions—in other words, the "stuff" of the atom is a *physical structure* in 3D time that is given a *coordinate location* (a point) in our observable, coordinate grid of space. The *location* is in space (yang), the *structure* is in time (yin).[3] This also tells us something important: we consider atoms to be *solids* (in various states) and *atoms are time*, so therefore **time appears as a solid**—and that is the stuff of *æther*—the "solid of time." All of our material particles, atoms and molecules are basically little balls of solid æther stuck on a 3D, empty spatial grid, exactly as the 19th century researchers said.

Now consider the reciprocal perspective. Larson agrees with Eddington in that everything that *exists in space also exists in time* and everything that *exists in time also exists in space*. And that includes three dimensions of space, three dimensions of time and clocks: *clock time* and *clock space*. Observation tells us we have locations in space and structure in time, so the reciprocal must also exist: *locations in time* with corresponding *structure in space*. This is what the early researchers called the *ætheric* realm and what Larson calls the *cosmic sector*.[4]

This interpretation of "space as empty" and "time as full" allows for *both* the conventional, material perspective of "little balls of time in the vacuum of space" and the original cosmic/ætheric concept of "little bubbles of space floating around in the solid of time"[5] to happily *coexist*, creating a more complete view of *The Structure of the Physical Universe*.[6]

MATTER       ÆTHER (cosmic)

Constructive
Start with nothing and
increase to something
(atoms are balls)

**3D Space**
Yang Location

**3D Space**
Yin Structure

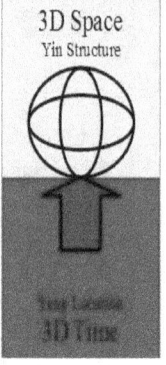

Destructive
Start with everything
and reduce to
something
(atoms are bubbles)

---

3    If you are familiar with the yin-yang model of the Universe, just substitute *space* for *yang* and *time* for *yin*, and you basically have Larson's theory. Once you get through the conceptual hurdles, it's very much the *Tao of Motion.* (Recent RS2 research now identifies *equivalent space*, the rotational projection of time into space, as *yin space*.)

4    As discussed in "Homo Sapiens Ethicus," the life unit is an aggregate of material and cosmic molecules, therefore having location and structure in *both* sectors.

5    Analogous to the Theosophical concept of "bubbles of koilon."

6    Larson, Dewey B., *The Structure of the Physical Universe*, North Pacific Publishers, Portland, OR, 1959.

**Motion, in a Universe of Motion**

With only space and time to work with and the definition of motion being a relation of space *to* time, there are *only two possibilities* that constitute "not motion": s/s and t/t.

When a motion with a net displacement in space interacts with another motion with a net displacement in time, they pass through each other because the relationship of s/t or t/s *constitutes motion*. But when two motions meet that have net displacements in the *same aspect*, they get stuck together because space-to-space and time-to-time *do not constitute* motion. This is how motions (particles, atoms and molecules) build up to bigger motions. *Chemistry* determines how atoms stick together, described by Larson in the latter half of *Nothing But Motion*[314] as just a process of *speed* and *orientation*—from which the concept of *valence* is derived.[315]

---

[314] Larson, Dewey B., *Nothing But Motion*, North Pacific Publishers, Portland, OR, 1979.

[315] *valence*, definition: "the quality that determines the number of atoms or groups with which any single atom or group will unite chemically." In the Reciprocal System, *all chemical bonds are the same*, using orientation to cancel net atomic speeds.

 In the Reciprocal System, chemical combination is a process of neutralizing speed by orienting atoms geometrically, much like assembling the pieces of a puzzle together—just need to arrange the pieces so the proper tabs and holes line up. The consequence is that *structure* in one aspect alters *location* in the other. We normally see just one side of this system, the material one, where structure in time (atomic properties) alters locations in space (physics and chemistry). The ætheric side follows the same rules, with the aspects of space and time reversed, but is interpreted more *metaphysically*—structure in space (Sacred Geometry) alters locations in time (electric, magnetic and bioenergetic fields). Just by understanding this "inverse chemistry," one can easily comprehend concepts such as *ley lines*[316] and *Feng shui*[317]—how spatial orientation affects energetic fields.

This can be demonstrated with the common electron moving through a conductor. In the Reciprocal System of theory, the electron is a "rotating unit of space." As such, it has its *structure in space* and its *location in time*—surprise, surprise—they got it backwards again! The electron is *æther*, not matter![318]

All material atoms are rotations in time, so an electron can pass through the time of the atom (space-totime constitutes motion) and we call that *electric current*, the flow of little bubbles of space in the solid, temporal atomic rotation of atoms—conventionally called a flow of "holes" in electronics (one of the few concepts that isn't backwards).[319] Because of this ætheric structure, we should see some modification of coordinate time—and we do, the resulting *electromagnetic field* when current flows through a wire and the electromagnetic field, in turn, will affect coordinate space rearranging objects along "lines of force." Once you understand that there are two halves (sectors) of the universe that are constantly interacting

---

[316] Watkins, Alfred Watkins (1925). *The old straight track: its mounds, beacons, moats, sites, and mark stones.* Methuen & Co Ltd. "Ley lines are supposed alignments of numerous places of geographical and historical interest, such as ancient monuments and megaliths, natural ridge-tops and water-fords. The phrase was coined in 1921 by the amateur archaeologist Alfred Watkins, in his books *Early British Trackways* and *The Old Straight Track*. He sought to identify ancient trackways in the British landscape. Watkins later developed theories that these alignments were created for ease of overland trekking by line-of-sight navigation during neolithic times, and had persisted in the landscape over millennia."

[317] *Feng Shui* (wind-water) is a Chinese philosophical system of harmonizing with the surrounding environment.

[318] The *positron* is the rotating unit of time, the "matter." The reason that electrons are abundant and positrons are rare in the environment is simple: the space of the electron passes *through* the time of atoms (space-to-time is motion), whereas the time of positrons gets stuck in the time of the atoms (time-to-time is not motion). Atoms absorb positrons, but allow electrons to pass.

[319] Though they do put out a lot of effort into making things backwards, as P.A.M. Dirac said, "an ordinary electron 'rests' on the Dirac sea, whereas a positron exists as a 'hole' in that sea."

using the *laws of **motion***, not *laws of **matter***, it becomes quite easy to understand how everything affects everything else.

Most of the interactions we are familiar with are mechanical or electronic, which are 1-dimensional interactions with 3-dimensional structures. Mechanical systems work with vectors and electronics work with the 1-dimensional rotational system of electrons. Life, Larson's "life units" as described in *Beyond Space and Time*,[320] are an interaction of a 3-dimensional material structure with a 3-dimensional cosmic structure; our *body is matter* and our *soul is æther*.[321] And that is Larson's definition of *life*: a stable (harmonious or sympathetic) combination of a material aggregate with a cosmic aggregate. Life is essentially a *stable* matter-antimatter reaction.

As mentioned in "Geochronology,"[15] when a star goes supernova the core is accelerated to faster-thanlight speeds, creating cosmic (ætheric) matter. The supernova remnants, such as white dwarf stars, planets and moons, are a combination of 3D matter **and** 3D æther.

Therefore, by Larson's *definition*, planets are *alive* and subject to the *laws of life*. Say "hello" to *Gaia*[16], or perhaps I should rephrase that to, "hollow" to Gaia—and here's why...

**Hollow Down There!**[322]

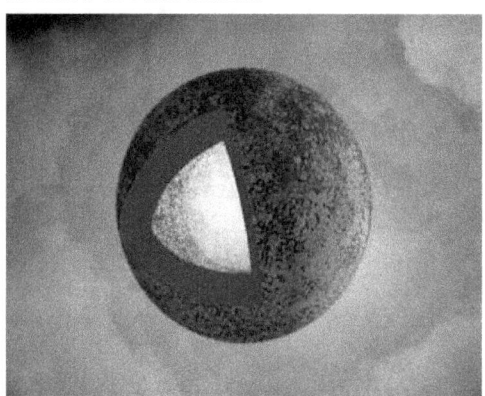

*"There are invisible worlds besides those perceived by us in our planetary system, unreachable centers of ethereal structure about us that stand in a higher plane of development than earthy matter, which is a gross form of disturbed energy."*[323]

One of the many natural consequences of faster-than-light motion in the Reciprocal System is the conclusion that FTL motion does *not* "time travel" to the past or future, but causes *inversion* of physical structure.[324] The moons and planets in *stable* orbits, being composed of a similar "inverse density"

*Figure 1: Cutaway View of Objects*       structure to white dwarf stars, have led to the

---

[320] Larson, Dewey B., *Beyond Space and Time*, Tucek & Tucek, 1995.

[321] The cosmic, 3D time structure of the soul is the origin of the concepts surrounding the *ætheric body*. 15 Phoenix III, Daniel, "Geochronology," page 9, "The Early Structure of the Solar System." 16 The personification of the Earth.

[322] Word play on the 1969 film, "Hello Down There," which is the story of a family living in a prototype underwater house.

[323] Drury, Llewellyn, *Etidorhpa or The End of Earth. The Strange History of a Mysterious Being and The Account of a Remarkable Journey*, 8th Edition, The Robert Clarke Company, Cincinnati, 1897, p. 76.

[324] Larson, Dewey B., "The Density Gradient in White Dwarf Stars" discusses how the imploding core of a supernova explodes in *time*, causing contracting and inversion of density in space.

conclusion that

*with FTL Cores* they are *hollow at the center* (lowest density) with a dense, protective surface.[325]

The concept is simple enough; those points of light in the heavens have a luminous surface with a dark, dense, temporal *core*.[326] Take its geometric reciprocal by yanking it inside-out and you get a dark, dense *surface* with a luminous core. In astronomy, the former is called a *star* and the latter, a *planet*.

Historically, the discovery of this "inverse density gradient" of the Earth was by unrelated means, from seismic soundings of earthquakes, deep mining, esoteric knowledge or metaphysical research that has been historically documented as the *Hollow Earth theory*. Larson's astronomical structure, as described in *The Universe of Motion*[327] not only indicates that the Earth is hollow, but *all* of the other planets, moons and dwarf stars follow the same pattern: a *hollow **planet*** theory. Because the "hollow planet" is a *natural consequence* of the Reciprocal System, you do not need people channeling Extraterrestrials nor Admiral Byrd writing a diary as he flies into a polar opening to describe it. The structure of the interior can be logically *deduced* in enormous detail.

The formation of white dwarfs, planets and moons are a consequence of a stellar or planetary explosion with sufficient energy to blow off the surface layers in an explosion in space, while concurrently imploding the heavier elements in the core, creating an *explosion* in *time*. What we see in space is a cloud of debris and a small, hard core that initially *looks* like a "black hole," though it is just matter moving at superluminal velocities emitting X-rays and gamma rays, as described by Larson in his paper on "Astronomical X-Ray Sources."[328] An interesting point here is that the black hole, being matter in the ultra-high speed range, will exhibit *anti-gravity motion* in space—not sucking in all the matter around it like the drain in a cosmic sink, but *repelling* matter (after all, that's what *anti*-gravity does). So what is seen in telescopes is an invisible (faster-than-light) source of intense gravity, that has pushed all the debris around it out into a ring,[329] misunderstood as an *accretion disk*. Because this "black hole" is still a material structure,[330] the temporal motion will eventually degrade increasing in visible, infrared and radio emissions, changing it

---

[325] Bodies without orbits, or in elliptical orbits, are simply inanimate and wholly material.

[326] Coronal holes, "… regions where the sun's corona is dark" (NASA) are indicative of this.

[327] Larson, Dewey B., *The Universe of Motion*, North Pacific Publishers, Portland, OR, 1988.

[328] Larson, Dewey B., "Astronomical X-Ray Sources," ISUS, Inc., Salt Lake City, UT, 1974.

[329] Motion in 3D time is observed in 2D "equivalent space" like a shadow, tending to take a planar geometry, such as a ring.

[330] Stellar structures that have sufficient temporal velocity to move into the cosmic sector are called *quasars*, not black holes. Once motion crosses that unit speed boundary, the black hole becomes its reciprocal: a white hole, which will eventually disappear from observation as temporal gravity pulls its structure into the unobserved, ætheric region.

into a brown dwarf. Give it enough clock time, that black hole will simply cool down to the white dwarf, then starts heating up to join the main sequence of stars.

### Outsides are In; Insides are Out

Faster-than-light speeds are *motion in time*, taking place in the cosmic sector or ætheric realm. What we find is that planetary bodies have *spatial shells* and *ætheric cores*, their version of a "body" and "soul," unless you happen to be standing on the ætheric core—then the shell is ætheric and the *core* is *spatial*. That's the tricky bit to understanding geometric inversion in four dimensions[331]—points and volumes swap places, so when making the transition from space to time, things get yanked inside-out.

Consider the atom described earlier, a volume of time located at a point in space. Now apply that to a *planet*—the Earth is a "volume of time" located at a "point in space," our orbit. The core of the planet would then be a volume of space—the electric, magnetic and gravitational "fields" located at a point in time, with a "temporal orbit" around the ætheric core of the sun. This is also why planets have stable orbits. Gravity is *inward in space*, but the *cosmic gravity* of the ætheric core is *inward in time* and due to the reciprocal relation of space to time, must be *outward in space* as *antigravity*. It is this push-pull arrangement that locks a planet in orbit. All stars and supernova fragments have this inside-out, outside-in structure to them.

Do you remember that cosmic structures (misnomered as "antimatter") are basically "bubbles" in the solid of time? Like positive "holes" in a conductor? It is not easy to observe a bubble *inside* something solid, as the solid stuff tends to block direct observation. That's why we have X-ray machines at the doctor's office. Cosmic structures are, therefore, *invisible* to material observers, because they are usually hidden inside something solid and we have no physical senses to directly determine lengths, areas and volumes *in time*.[332]

Applying the "bubble" concept to a large aggregate such as the Earth, what we end up with is a hard, spatial shell with a hollow, empty, bubble-like core that just happens to appear super-dense,[333] because it is moving in *coordinate time*, not space. This is the conceptual origin of the *Hollow Earth* theory.

---

[331] I had put this at "three dimensions," but that is not technically correct. In the material sector, there are three dimensions of space and one of time (clock time), requiring four independent variables to express—a 4D system.

[332] But we do have "nonphysical senses," known as extra-sensory perception, ESP, or in the broad sense, *psionics*.

[333] Density is determined by net, temporal displacement. As such, coordinate time, 3-dimensional time, looks like the *cube* of normal 1-dimensional time, making it appear *extremely* dense. $10^1 = 10$, whereas $10^3 = 1000$.

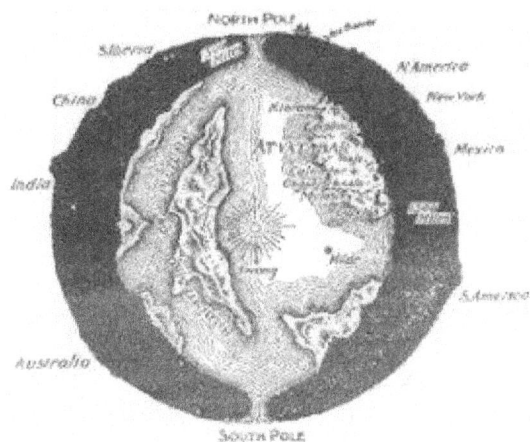

The picture is from *The Goddess of Atvatabar*,[334] depicting this hollow Earth structure and, like most hollow Earth theories, includes a "central sun" to light up the convex, inner surface simply because no other mechanism could be fathomed at the time. Those people that have made the journey describe it as, for example, "The great luminous cloud or ball of dull-red fire—fiery-red in the mornings and evenings, and during the day giving off a beautiful white light, 'The Smoky God,'—is seemingly suspended in the center of the great vacuum 'within' the earth, and held to its place by the immutable law of gravitation, or a repellent atmospheric force, as the case may be."[335]

The rendering of a new planet with a faster-than-light core in Figure 1 shows a luminous center—not necessarily a "central sun" but more an "inverse photosphere." To explain how this works and why it is perceived as such, time to call up that reciprocal relation between space and time: the Reciprocal System of theory.

Dewey Larson and the researchers of the last 50+ years have not considered these concepts before, as they were busy promoting the physics and chemical aspects of the theory, with some astronomic papers addressing galaxies, quasars and pulsars. In 1998, "At the Earth's Core, The Geophysics of Planetary Evolution"[336] was published in the journal, *Reciprocity*, opening the door to understand what is going on "down under." It is now almost 20 years later and considerable progress has been made in understanding the structure of planets, moons and stars in this context.

### The Ins and Outs of Insides and Outsides

In order to understand just how planets, stars and moons can be hollow and how people can stand on the inner surface without falling into the center of the planet, we need to take a better look at the Reciprocal System concept of "speed ranges," which I've referred to on many occasions in prior papers.

    A       B       C       Larson defines three material speed ranges, normally used in

---

[334] Bradshaw, William R., *The Goddess of Atvatabar; Being the History of the Discovery of the Interior World, and Conquest of Atvatabar*, Arno Press, 1975.

[335] Emerson, Willis George, *The Smoky God or a Voyage to the Inner World*, Forbes & Col, Chicago, 1908.

[336] Peret, Bruce, "At the Earth's Core, the Geophysics of Planetary Evolution," *Reciprocity* 27 № 1 page 9; journal of the International Society of Unified Science.

astronomical context, where they are most apparent:

1-x: "Low speed," the conventional realm of a 3D, gravitational system.

2-x: "Intermediate speed" used to explain motions in equivalent space that seem to be immune to the effect of gravity.

3-x: "Ultra high" speeds that exhibit anti-gravity motion that moves opposite to the pull of gravity, such as pulsars.

But there is a problem with Larson's model—the second scalar [3-x] dimension (B) is split in half, meaning that it operates *differently* than the other two (A, C). Because all three scalar dimensions are *independent*, they must all be *identical*, having the same structure and properties.

The RS2 reevaluation found a solution to this problem that requires an understanding of a strange geometric concept called a *quaternion*. No, it is not a Chubby Checker dance—though it twists like one![337]

A quaternion is basically a complex number[338] in three dimensions. But first, let's fix a misunderstanding about "imaginary numbers" that was *deliberately introduced* to prevent researchers from putting math and equations to ætheric structures.

**Quaternions: Forbidden Knowledge of the Mathematicians[339]**

In the beginning of the 19th century, non-Euclidean geometries began to generate interest and the old, mathematical rules that had stood for nearly two thousand years were brought into question. The *complex number* $i = \sqrt{-1}$ had entered the field and upset the applecart of the mathematicians. It *looked* like a rotation, but mathematicians were unsure what to attribute it to, while physicists remained blissfully ignorant. Sir William Rowan Hamilton of Dublin (1805-1865), aware of the controversy over the complex numbers, set out to find a consistent algebra for these numbers. He realized that this algebra related to the physical concept of *time*, saying that this "Algebra… viewed not merely as Art or Language, but as the Science of pure Time." If one complex number generates a rotation, he figured, two complex numbers should cover all of 3D space. But, way back in 1843, he discovered that it needed *three* complex numbers, labeled: *i, j* and *k*, to function with 3D space. These complex numbers, corresponding to the three *rotational axes*, along with the *real number*, make up the *quaternion*, following these rules:

$$i^2 = j^2 = k^2 = i\,j\,k \quad i\,j = k \quad i\,j = -\,j\,i \quad q = \omega + \langle i x + j y + k z \rangle$$

---

[337] Hank Ballard and the Midnighters, "The Twist," a pop song popularized by singer Chubby Checker in 1960.

[338] A complex quantity is a pair of numbers with "real" and "imaginary" components. The real portion is the conventional numbering system, and the imaginary part is based on the equation: $i = \sqrt{-1}$, to which there is no "real" answer— hence, "imaginary." But note that all our electronic technology is based on the imaginary number, so it must have some real existence.

[339] This historical section on quaternions was provided by Prof. Gopi Krishna Vijaya, Ph. D. specifically for this paper.

Where ω was the real number, a *scalar* (magnitude only) and $<ix+jy+kz>$ was a 3dimensional *vector*. Hamilton was the first to introduce the terms "scalar" and "vector." He also introduced non-commutativity (the order in which mathematical operations were performed, *mattered*). This is sensible in terms of rotation, as rotating around X-axis and then along Y-axis is different from rotating first around Y and then along X.

To those who had no idea what *one* complex number meant, let alone three, this was scary. But this triplet proved quite useful and Maxwell incorporated them in his famous equations of electromagnetism. This brought them into the domain of physicists and caught the attention of Josiah Willard Gibbs (1839-1903) and Oliver Heaviside (1850-1925). Both of them, independently, tackled quaternions in Maxwell's works and decided to *remove* the complex nature of the numbers. Physics of the time had no rotations to map complex numbers to and, as a result, the physicists preferred the *linear* "real number" version.

Heaviside complained: "how can the square of a vector be negative?" So they *dropped the complex numbers* and forced the vector part of the quaternion into modern Vector Analysis or Vector Algebra, using rules like *cross product* (the "right hand rule" in electromagnetism). The scalar part was kept aside, with rules relating to the *dot product*. The quaternion was broken into two convenient pieces: *scalars* and *vectors*.

Hamilton's supporters were not going to accept this dismemberment without a fight, and their fight (see *A History of Vector Analysis*[340] by M. Crowe) involved eight scientific journals, twelve scientists, and roughly 36 publications between 1890 and 1894. After this, with the increased utility of the vector algebra, practical concerns won the day and quaternions were pushed out of the mainstream. Vector algebra that is still taught today got entrenched into the textbooks.

However, an idea whose time had come could not simply be squashed out of existence simply for convenience's sake. After a couple of decades, the notion of quaternion would again poke out in two different streams. One stream picked up the complex number again and incorporated it into a 4D space-time. This is what we now know as *Special Relativity*. Another stream picked up the non-commutativity as well, giving rise to *Quantum Mechanics*.

Paul Dirac, one of the pioneers of this subject, was fascinated by Hamilton's work and even introduced the Hamiltonian equation into quantum mechanics. Quaternions was resurrected again, as were complex numbers, but without a clear connection to their history. All the troubles in understanding quantum mechanics to this day stem from the properties of the complex number and non-commutativity of quaternions, the same thing Hamilton was tackling two centuries ago. Both mathematicians and physicists have been at a loss to explain how physical quantities can be "imaginary," where the rotation called "spin" comes from and how the *order* of physical measurement matters.

Since imaginary numbers *cannot be directly represented* on the real number line, *nonlocality* was introduced into physics, which was another hard pill to swallow.

---

[340] Crowe, Michael J., *A History of Vector Analysis*, Dover Publications, Inc., New York, 1985.
ISBN: 0486649555, ISBN-13: 9780486649559.

Understanding the quaternion as an *expression of rotation* hence not only clears up these problems, but clears the way after nearly two centuries of being lost in the woods.

## Imagine That

If you look up "imaginary numbers," you will quickly discover that they are treated the same way as "real" numbers, but placed on an orthogonal axis. That is because conventional science, like Larson's original work, is very much a linear, yang system and had no way to express the rotational, yin system that the imaginary quantity represents. At least Larson made an attempt at it by introducing his concept of a *rotational base*, which forms the base of all particles and atoms.[341] This resulted in Larson splitting his "dimension of scalar motion" into two *units of motion*—speed, a linear, *kinetic vector* and energy, *torque*, the energy of rotation— basically, how fast a ball is moving in a straight line, and how fast it is spinning as it moves. Larson's "two units" of a scalar dimension are actually a *complex* quantity: $Re$(speed) + $Im$(energy), though he never realized it.

In order to better understand the 3D quaternion, let's first look at the 1D "complex" version. The conventional Argand diagram with a *real* (x) and *imaginary* (y) axis is basically a comparison of apples to elephants, two different concepts trying to be expressed the same way. This added a great deal of confusion to what would otherwise be a very simple concept: that of *turning around*.

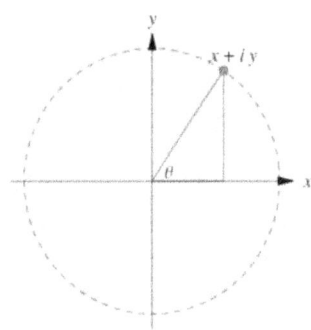

*Figure 2: Argand Diagram*

Linear geometry is *additive*, 2 feet + 3 feet = 5 feet and you stay on the same line of motion, but rotational geometry is *multiplicative*, 2 feet × 3 feet = 6 *square feet*. Or, in the "imaginary" sense, if you do a rotation of "$2i$," then rotate again by "$3i$,", you get $2i \times 3i = 6i^2 = -6$. You did not move linearly, but spun around an unrepresented axis that is sticking out of the page on the graph, ending up *not with an area*, but on the same axis pointing in the opposite direction. Imaginary numbers are better expressed as "rotational operators," usually represented by the letters $i$, $j$ and $k$, and serve the purpose of telling you that the associated magnitude refers to *rotation*, not *translation*.

The 3D version is actually quite simple—if you treat a *rotational operator* as a *rotation*. To accomplish a 3-dimensional rotation, try this example:

---

[341] This has been a point of contention in the Reciprocal System for decades, as Larson insisted that "you *can* have motion without something to move," but at the same time, *could not* have rotation, without something to rotate. The RS2 reevaluation addresses this problem by including "rotation without anything to rotate" as a natural consequence.

See: KVK, Nehru., "The Law of Conservation of Direction," *Reciprocity* XVIII, No. 3, page 3.

1.        Take your right arm and point forward. This front-back line, formed by your arm, is the "real" axis. Forward is +1 and behind you is -1. This is an arbitrary selection, as you can start in any direction and call it "forward" or the "real" axis. It is just the starting direction of rotation.

2.        Now, rotate your shoulder so your arm points straight up. This is a rotation of *i*, being made in the plane formed by the vertical and where your arm was pointing. Note that a rotation occurs *in a plane*, not up a line, about the axis that is orthogonal (perpendicular) to the plane of rotation.

3.        Rotate your arm so you now point to the right. That is a rotation of *j*, in the plane orthogonal to the real axis, so we are actually rotating *about* the real axis, going from pointing up, over to pointing right. Note that the *j* axis and real axis *are the same axis*—except when doing "real" motion, we are *sliding* your arm in and out, rather that *twisting* it in a rotational fashion.

4.        There is only one plane left that you have not  moved in, the horizontal one. Take your arm and rotate it around horizontally until you are pointing backwards. You are now pointing at "-1" on the real axis. This is a 3D, quaternion rotation. Also note that your hand is upside-down from where it started.

Your arm is *always* rotating in a plane, not moving up or down an axis, as the Argand diagram does. Think of it in these terms and you will start to understand the rotational, vortex-like action of æther.

An easy way to remember how imaginary numbers (rotational operators) work, is to think in terms of how an airplane, submarine or spaceship turns: *roll, pitch* and *yaw*. That's the "imaginary vector" of the quaternion, where the line of flight is the "real" component.

**Revisiting Speed Ranges**

With an understanding of the quaternion as a 4-dimensional quantity, consisting of one real and three rotational operators, we can apply that knowledge to what Larson was trying to express with his "speed ranges," switching between speed (translation) and energy (rotation).

As discussed in prior papers, the *natural datum* of the Reciprocal System is *unity*. Everything starts with *one* (unit, outward speed), and "displaces" away from it. When we look at moving through a 3dimensional, quaternion rotation, one finds that it parallels Larson's three speed ranges, with the 1-x being the *i* rotation, 2-x being the *i.j*, and 3-x being the *i.j.k* = -1 (inward speed).

| Speed Range | Rotation [ω <i j k>] | Function |
|---|---|---|
| 1 | 1 0 0 0 | Outward (+1) progression of the natural reference system known as the *Hubble expansion*. |
| 1-x *low* | 1 i 0 0 | Our conventional reference frame of the 1-dimensional, *dielectric* rotation described by vectors and spins. |
| 2-x *intermediate* | 1 i j 0 | Since i.j = k, appears as a 1dimensional motion but with 2D properties—*electromagnetism*. |
| 3-x *ultra high* | 1 i j k | Since i.j.k = -1, net inward motion in 3 dimensions: *gravitation*. |

By doing this 3D "twist," we now have the equivalent of Larson's *direction reversal*, an inward motion that can counter the outward progression to create a tangible object. But it goes beyond that.

### The Structure of the Living Planet

Up to now, the structure of the planet has always been considered to be *inanimate* by geophysicists and the like. But in Larson's posthumous publication of *Beyond Space and Time*, he revealed the true nature of life as a *stable combination of material and cosmic rotations*, a "matter-antimatter" reaction that *implodes* to produce living energy, *bioenergy*. Not long thereafter, it was recognized that life behaved much the same as *intermediate speed* motion, which was normally associated only with astronomical objects.

We already know that planets are inside-out stars, with a 3-dimensional, *spatial* exterior and a 3dimensional, *temporal* interior. In order to get from one to the other, we must move not only through *all three* speed ranges of the material, spatial half of the structure, but also the three cosmic speed ranges, to reach to the opposite extreme and get a complete picture of what a planet actually looks like.

First, we need to identify these "opposite extremes." In the Reciprocal System, everything is based on *discrete units* or *quanta*. What that means is that the *smallest quantity* you can have of anything is *one*. You cannot have "none" of anything (0), nor can the universe "owe you one" (-1). You either have one or more of something or you do not have it at all.

We must consider that *gravity* works that way. As we move away from the Earth, we reach a point where we *no longer have* at least one unit of inward motion (gravity) and the pull of gravity just *disappears*, completely. It doesn't slowly fade away, it just *stops*. That is termed the *gravitational limit*.

Without gravity, we also do not have a 3D, spatial coordinate system, because we no longer have a *center* to pull things towards, to give us some kind of orientation as to what is up and down. Outside the gravitational limit, everything *progresses*—it just wants to just fly apart at the speed of light. This is observed as the Hubble expansion, the scalar expansion of the Universe by astronomers. Yet, when they are taking photos of distant objects, they never consider that space is acting *differently* between "galaxies."[342] They just assume it works exactly the same way it does in their living room.

But, what is particularly interesting about the gravitational limit, is that it is also the place that allows the *transition* from motion in space to motion in *time*. Space/time is a fraction of the speed of light, $1/n$, the speed of light is unity, $1/1$, so time/space, the cosmic, ætheric realm, must be $n/1$. To go from $1/n$ to $n/1$ requires passing through $1/1$, the gravitational limit, which acts like a "light speed barrier" much like a "sound barrier."[343]

For our world, the outer gravitational limit is just outside the orbit of the Moon, about 300,000 miles away. This forms an impenetrable barrier[344] to our technology, because our rockets are based on *vectorial* motion through coordinate space. Past the gravitational limit, there is *no* coordinate space, so rocket engines just don't do anything (conventional laws of physics do not apply in a scalar zone). Why is this important? Let's take the flip side—we know that the temporal core of our planet is moving in time, not space, so somewhere inside the Earth there must be an "inner" version of the gravitational limit, to allow that transition between motion in space and motion in time to occur. And when we "dig deep," that is what we find inside the structure of the Earth.

I would like to note that when it comes to space travel, *Science is Fiction.*[345] All those wonderful photos that the United States space agency, *Never A Straight Answer* (NASA) has been publishing are just *computer simulations* of telescopic images. Their reasoning is quite logical—if *we* can't get there, *you* certainly aren't going to, so you can never prove these are fake and we'll just put on a good show to get lots of money from the ignorant masses. What a wonderful world we live in.

### The Material Aspect of Tiamat: Earth

Let's take a detailed look at the structure of Tiamat, our watery world, to see how

---

[342] Galaxies are actually newly forming solar systems. See: "The Colonization of Tiamat (Part IV)" for details.

[343] The sound barrier or sonic barrier is a popular term for the sudden increase in aerodynamic drag and other effects experienced by an aircraft or other object when it approaches supersonic speed. When aircraft first began to be able to reach close to supersonic speed, these effects were seen as constituting a barrier making supersonic speed very difficult or impossible.

[344] Referred to as the "stargate" by Michael Tsarion, in his lectures. The New World Order mistakenly assumes that this *natural* barrier is some kind of quarantine field created by extraterrestrials, powered by technology on the moon. Sorry, NWO, but your energy-based weapons won't make a dent, because your EM weapons that operate at the speed of light, cannot pass a light speed barrier.

[345] Phoenix III, Daniel, "Science is Fiction," July 6, 2015 blog post on http://conscioushugs.com

these pieces fall into place. With that, we can make some mythological connections and identify the ancient worlds.

Humanity knows very little about what is underneath the ground, mainly because it has been covered up with all kinds of dirt, for centuries. The Powers That Be would have quite a mess on their hands if the general muggle population were to become aware of the facts that "we are not alone," and it isn't as much from "out there" as "in here," right beneath our feet.

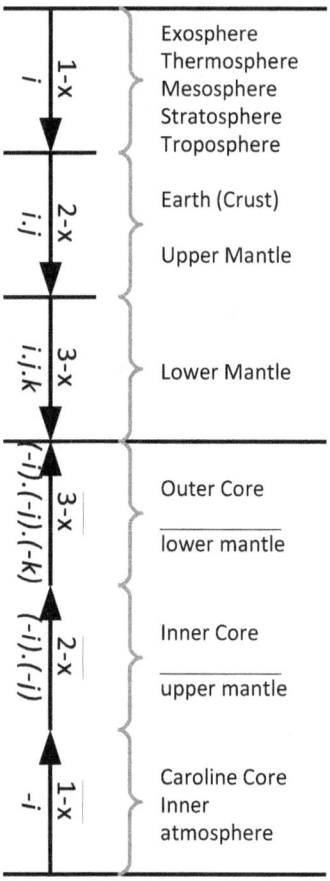

+1But before we get beneath our feet, let's put our heads in the clouds

and start at the outer, gravitational limit and work down, following the pull of gravity.

Using our new understanding, we find that the low speed range (1-x) is primarily 1-dimensional in nature, composed of electric charge and thermal motion. If we look to the sky, we find just that. Firstly, is the *exosphere*, a region of charged particles followed closely by the *thermosphere*, "hot stuff" with metal-melting temperatures reaching 2000 °C (3,630 °F).

As we approach the surface, we run in to the *ionosphere*, where the electric effects meet interference from the now nearby intermediate

-1speed range (2-x) of the crust. Below that, the *mesosphere, stratosphere* and *troposphere* form the gravitationally-bound pile of gaseous atoms we call our atmosphere—but still exhibit strong, electrical effects, such as lightning, cosmic rays, elves and sprites.[346]

The transition from low to intermediate speeds is about 30 miles beneath our feet, at the junction between the crust to the upper mantle. Here we see the transition from primarily charge and thermal motion, to electromagnetic motion.

As mentioned earlier, the intermediate speed range is the *region of life*, and what do we find just above the surface, on the surface and immediately below it? Birds, trees, bugs, people, fish and burrowing

+1critters—the region is *loaded* with life, as expected.

But it does not stop there, as the intermediate speed region extends down about 250 miles, which means *life* also extends at least that far. There are records from all over the world, documenting all manner of cave-dwelling life, from Dwarfs, Goblins, Kobolds, Knockers, Abandondero, Tero, Nagas,... to the Gods,

---

[346] Elves and sprites are a form of upper-atmospheric lightning (or ionospheric lightning), a family of short-lived electricalbreakdown phenomena that occur well above the altitudes of normal lightning and storm clouds.

themselves, with their huge, underground empires. Why would so many of these stories exists in every ancient culture across the planet, if there was not an element of truth to them?

As we move into the ultra-high speed range (3-x), rock starts to thin out because of incompatibility with the approaching "center of gravity," much like the atmosphere begins to thin out with altitude. When we reach the bottom of this limit, indicated as "-1" on the graph, solid rock has become more gaseous in nature, creating a zone of mist.[347] It is this "inverse gravitational limit" where all things come to rest, and, being spherical in shape at the lower mantle / outer core boundary, is referred to as "the sphere of rest" in ancient literature. It is at this point that the push of gravity from above meets the push of gravity from below— canceling each other out, resulting in a zone of weightlessness. As described in *Etidorhpa or The End of Earth, The Strange History of a Mysterious Being and The Account of a Remarkable Journey*:[348]

"You are to proceed to the Sphere of Rest with me," he replied, "and in safety. Beyond that an Unknown Country lies, into which I have never ventured."

"You speak in enigmas; what is this Sphere of Rest? Where is it?"

"Your eyes have never seen anything similar; human philosophy has no conception of it, and I can not describe it," he said. "It is located in the body of the earth, and we will meet it about one thousand miles beyond the North Pole."

…

"At another time, perhaps," he remarked; "we have reached the Inner Circle, the Sphere of Rest, the line of gravity, and now our bodies have no weight; at this point we begin to move with decreased speed, we will soon come to a quiescent condition, a state of rest, and then start back on our rebound."

"If you will reflect upon the condition we are now in, you will perceive that it must be one of unusual scientific interest. If you imagine a body at rest, in an intangible medium, and not in contact with a gas or any substance capable of creating friction, that body by the prevailing theory of matter and motion, unless disturbed by an impulse from without, would remain forever at absolute rest. We now occupy such a position. In whatever direction we may now be situated, it seems to us that we are upright. We are absolutely without weight, and in a perfectly frictionless medium. Should an inanimate body begin to revolve here, it would continue that motion forever. If our equilibrium should now be disturbed, and we should begin to move in a direction coinciding with the plane in which we are at rest, we would continue moving with the same rapidity in that direction until our course was arrested by some opposing object. We are not subject to attraction of matter, for at this place gravitation robs matter of its gravity, and has no influence on extraneous substances. We are now in the center of gravitation, the 'Sphere of Rest.'"

"I am the man" and his Guide may not have known what lies beyond, but it actually isn't difficult to figure out, as it has been heavily documented by the

---

[347] Yes, this is the same mist described in my paper, *Homo Sapiens Ethicus*, as part of the death/dying process. A soul can make this journey, as well as a body—but only if you are on your way to Hell.

[348] Drury, Llewellyn & Lloyd, John Uri, *Etidorhpa, op. cit.*, pp. 129, 321, 340.

Romans, Greeks, Vikings, Mayans, Chinese and other cultures—complete with detailed maps and descriptions—if you know *where* to look, and what you are looking at.

First, let's continue with a structural analysis, continuing "downward" from the Sphere of Rest, into the conjugate realm of the outer core—and inner earth.

The Sphere of Rest, being the center of gravitational attraction, means that everything *above* is pulled *downward* towards the Sphere, while everything *below* is pulled *upward* towards it. Our surface world is convex and centripetal in nature, whereas the inner realm is concave and centrifugal in nature.

As the saying goes, "as above, so below," and that is the case here—the region from the Sphere of Rest to the inner gravitational limit (the "+1" at the bottom of the graph) has the same structure as the region from which we just traversed, up to the outer gravitational limit (the top "+1"). The inner and outer worlds are like mirror images of each other, with the surface of the mirror being the Sphere of Rest. But there are some interesting differences, as this inner realm is overlapped at a macrocosmic "intermediate speed range" with 3D time. This inner realm, as we shall see, tends to be the destination of souls on their reincarnative journey, as mentioned in *Homo Sapiens Ethicus*.[349]

### The Body and Soul of Planets

Our investigation, so far, has dealt only with the material "body" aspect of the planet, the "Earthly" the aspect we are familiar with. We found it consists of two, quaternion-like structures, each having three, rotational speed ranges with distinct properties. These "dual quaternion" structures form a dimensional relationship that is analogous to Larson's two "units of motion," speed and energy.

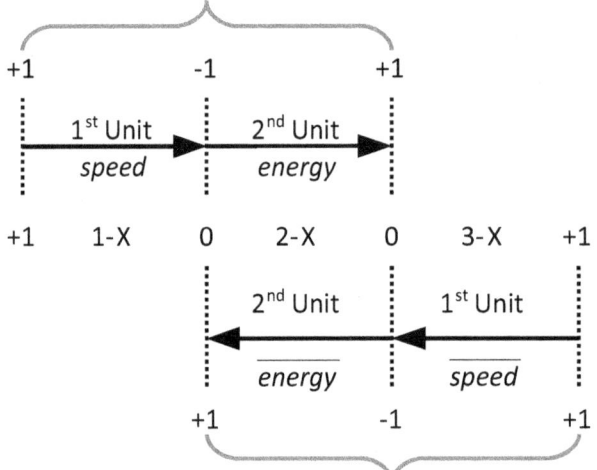

Material: 3D Space

Earth

In order for the cosmic sector to link to the material, a planetary "silver cord"[350] must exist, tying the two together, working in the same fashion as chemistry does with molecules: arrange the puzzle pieces so that they click together, with a zone of zero, net speed at the interface. That way, with no net motion between Cosmic: 3D Time them,

---

[349] Phoenix III, Daniel, "Homo Sapiens Ethicus," available on
http://conscioushugs.com or http://reciprocalsystem.org

[350] "The silver cord in metaphysical studies and literature, also known as the sutratma or life thread of the antahkarana, refers to a life-giving linkage from the higher self (atma) down to the physical body." https://en.wikipedia.org/wiki/Silver_cord

the pieces stay attached.

*Agartha* Figure 3 shows how the two sectors of the
*Figure 3: Inter-Sector Bonding* Universe, the material and cosmic, link together to form that zone of zero, net motion zone (2-x),
which just happens to coincide with the region of the "inner earth"—and also, the reciprocal inner world, from the region of 3D time.

### The Destination of Souls… What the Hell? Yes.

In metaphysics, the expression of consciousness is split into three complexes: the body (*corpus*), the mind (*anima*, or soul), and spirit (*animus*).[14] Planets, being a life form, follow the same pattern—just on a larger scale.

The first unit of motion in the material sector is our earthly existence, that of the sky, ground, water and the depths of the ocean and underground. Very materialistic for a material realm, because materialism is a natural bias here.

The second unit of motion has the same structure, but sits back-to-back against the Sphere of Rest, forming an inner surface with an inverse gravitational pull. You stand in the inner realm with your head pointed at the center of the planet, not towards the stars!

This realm is known in the ancient records as the *Underworld*, and also by a few other names, like:

Ferri, Mictlan, Irkalla, Naraka, Annwy, Mag Mell, Diyu, Hel, Hell, Sheol, Aaru, Toonela, Elysium,

Echeide, Guayota, Yamaloka, Patala, Maski, Alvilág, Uku Pacha, Adlivun, Jahannam, Naar, Barzakh, Araf, Adho Loka, Yomi, Ne-no-Kuni, Jigoku, Ji-Ok, Aizsaule, Alam Ghaib, Hawaiki, Rarohenga, Pellumawida, Degin, Wenuleufu, Ngullchenmaiwe, Metnal, Xibalba, Bulu, Burotu, Murimuria, Nabagatai, Tuma, Gimlé, Niflhel, Vingólf, Ekera, Duzakh, Kasanaan, Avaiki, Bulotu, Iva, Lua-o-Milu,

Nga-Atua, Pulotu, Rangi Tuarea, Te Toi-o-nga-Ragna, Uranga-o-Te-Ra, Shipap, Inferno, Avernus, Tărâmul Celălalt, Nav, Podsvetie, Peklo, Iriy, Dilmun, Kur, Irkalla, Hubur, Erlik, Guinee and Hiyoyoa.

I am going to use the Greek name *Hades* in this paper, because of all the names, it has the least religious context, is fairly familiar and clearly identifies the realm under discussion. So where in Hell does this leave us? Right about here, just down the street from Knockturn Alley:[351]

---

[351] From the Harry Potter series, Knockturn Alley (a play on the word "nocturnally") is a dark and seedy alleyway leading off from the more savoury Diagon Alley to which Muggles have no access. It is frequented largely by Dark Wizards.

This is a section of the map of Hades, detailing some of the major, after-death attractions. In ancient times, Hades was entered via the Cape Matapan Caves, located on the southernmost tip of Greece. Not too accessible these days, as ocean water levels have risen substantially in the last 500 years. Other cultures document additional human-accessible routes via the volcano Hekla, in Fjallabaksleið Syðri, Iceland, Fengdu City in Chongqing, China, Lacus Curtius in Rome, Italy, Actun Tunichil Muknal, in the Tapir Mountain Nature Reserve, Belize, The Gates of Guinee in New Orleans, USA, Pluto's Gate in Denizli Province, Turkey, St. Patrick's Purgatory in Lough Derg, Ireland, Chinoike Jigoku in Beppu City, Japan, The Seven Gates of Hell in Hellam Township, Pennsylvania, USA, and the most famous: the Cave of the Sibyl in Naples, Italy.[352] Other known spots include the Death Valley in California and Cusco, Peru (which was still quite active during the Spanish conquest of the Americas).

My point to listing all of these references is that if Hades was just "fantasy," why does the Underworld exist in *every* ancient culture on the planet, with virtually the same description? It's not like the Roman Empire had a PC with an Internet connection in every hovel, doing global podcasts on their religious beliefs. In ancient times, communication and the exchange of information was difficult, at best, and restricted to the very few people that could read and write. The reason

---

[352] From the *Atlas Obscura*, http://www.atlasobscura.com/articles/the-atlas-obscura-guide-to-gateways-to-hell 48 Hesiod, *Theogony* 720–725 BCE.

all of these references exist is because they are all talking about a *real*, tangible place, what we call the "inner world" of hollow Earth fame.

Hades is the region in Figure 3 on the Material side, marked "2-X." If this diagram is accurate, then there should be *another level* to Hades, across the unit speed boundary of the inner, gravitational limit, making the place virtually inaccessible—another barrier—for the same reasons that the outer gravitational limit acts like a barrier to keep humanity quarantined to this world. Lo and behold, the upper left of the map of Hades shows exactly such a place, blocked off by a deep abyss: *Tartarus.*

Tartarus exists, as ENLIL (Zeus) commented, "as far beneath Hades as heaven is high above the Earth" and as Hesiod commented, "if a bronze anvil falling from heaven would fall nine days before it reached the earth, the anvil would take nine more days to fall from earth to Tartarus."[48] This is demonstrating the reciprocal relation that exists between the limits: 9 days up to the outer limit and 9 days down to the inner limit. $1/9 \leftarrow 1/1 \rightarrow 9/1$.

So here we have two different aspects of the Underworld, Hades on the material side and Tartarus on the cosmic side, separated by a virtually impenetrable, natural barrier. So what does this barrier look like, and how does it work? Easy enough—it works *almost* the same as the outer, gravitational limit, but because we are dealing with the "innards" of a living organism, the inner gravitational limit has an overlap of motion in space with motion in time. This overlap, being in the inverse low speed range is basically a 1-dimensional, simple rotation.

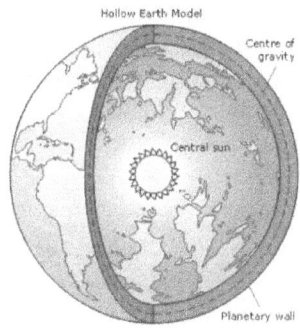

In the Reciprocal System, those simple rotations form the basis of the *positron* (1D temporal rotation) and *electron* (1D spatial rotation). When they mix, they form a *birotation,*[353] which, through the process of dimensional reduction known as Euler's formula,[354] ends up being a *cosine wave*—electromagnetic radiation. This boundary separating Hades from Tartarus will glow with light and heat: the *central sun.*

Curiously enough, there actually is a diurnal cycle to the light emitted from this internal sun, because its activity is connected to the external sun by lineof-sight intra-atomic (ætheric) energy transmission. As the Earth rotates, this "silver cord" connecting the central sun to the outer sun passes through different densities of material, sometimes land masses, other times, oceans. "Daytime" occurs when the line is passing through the Pacific Ocean, because water offers the least distortion to the transmitted energy, giving the internal sun a white, full-spectrum appearance. "Nighttime"

---

[353] KVK, Nehru, "On the Nature of Rotation and Birotation," *Reciprocity*, Vol. XX, No. 1, p. 8.

[354] A formula that shows a fundamental relationship between trigonometric functions and complex numbers: $e^{ix} = \cos(x) + i\sin(x)$. When two opposite rotations are involved, the formula reduces to the cosine wave function.
See: https://en.wikipedia.org/wiki/Euler's_formula

occurs when that line passes across continents, Eurasia in particular, where the 30 mile layer of rocks and minerals distort this ætheric communication, causing the central sun to "fuzz out" a bit and cool down towards the red/orange part of the spectrum.[30] Because of this relationship, night in Europe is daytime in Hades. This exchange of day and night between the surface and inner world has been documented by travelers to that realm, mistakenly assuming that the relation was based on the fact that Lucifer, being the rebel, deliberately switched them around—just to irritate God. As it turns out, this circadian correlation is a natural consequence of planetary structure.

### Summary of the Material Aspect of the World

From our material side, moving through the speed ranges gives us the atmosphere, the surface world, the underworld and the central sun, exactly as depicted by those that have ventured to those regions.

The Sphere of Rest, which pulls everything towards it, is also the zone of neutral gravity, where inward pulls cancel each other out.

The central sun is a hollow shell that delineates motion in space from motion in time with the glow resulting from the interaction between the two realms (just as our sun is also a hollow shell).

As disclosed in prior papers, we know that our planet, like all planets, expands over clock time, with the cracks filling in with water and evolving into oceans. That has some consequences for Hades. Back in the days of the ANNUNA colonization, the seas were very small, which means that the central sun tended to stay dull and reddish-orange in color. This is what gave the early depictions of Hell as a hot, dismal place of dull red fire—because it was. Recall that humanity was created *specifically* to be *slaves to the gods* and were forced to work in these regions, under these conditions. As a consequence, that image of Hades is buried deep in the collective unconscious of all humanity. But… is it still true?

I'm sure this will come as a big surprise, but *no*, it is not.

We have a substantial amount of surface area that is now ocean, allowing the central sun to brighten up towards full spectrum. This is literally turning Hades in to a Paradise. Even in Greek times it is mentioned that Hades was covered with fields of asphodel, a beautiful white flower, and the reddish skies are turning blue as a consequence of water vapor from the developing, inner oceans (which are

*Asphodel* meteoric deposits covering its surface that degenerate to salts). If humanity had direct access, the Blessed Isles would be full of pricey hotels and condos by now!

Let me conclude this summary with something for your consideration: Hades is, by definition, "where souls go after death." The soul, being the cosmic aspect of human life, is "antimatter," a 3D, spatial rotation (structure) and can easily move through the atomic, temporal rotations that compose the crust, oceans, and mantle of the material aspect of our world, since space-to-time constitutes motion. All that mythos regarding a soul's journey to Hades is basically *true*, because it has the proper space-time construct to move downward to the inner earth.

However, the soul *cannot* travel across space into heaven, because space-to-space is *not* motion. But the body *can*, hence "the resurrection of the body" concepts put forth by Christianity. Your body, being a temporal displacement, *can* traverse the vacuum of space—but your soul, a spatial displacement, *cannot*. For someone to actually reach heaven, they would then have to be purely materialistic bodies, with little to no soul to impede the way. Say "Hello" to *Royalty*, the soul-less, materialistic rulers of the world. *They know what they are doing.*

Consider that if you want to assist a soul on its journey you should put the body in contact with the ground, creating an easy path for the soul to follow and get started on its trip. This is the reason why good people are *buried*. However, to punish a soul and prevent that journey, one would have to prevent the contact between the body and ground—burn the witch! Many of these rituals have a scientific basis, once you understand the premises behind them.

*Everything you know is backwards*, including most religious belief. When you die, you go to Hades as a natural consequence of the structure of the soul. From there, you can proceed elsewhere, towards reincarnation, ascension, or to the Other Realm, the "flip side" of the surface world of the Earth, the region existing in 3-dimensional time inside the central sun, known to its inhabitants as *Agartha*.

### The Cosmic Aspect of Tiamat: Agartha

So far, we have covered the region from the gravitational limit, just outside the orbit of the moon, down to the central sun and just over its border to Tartarus. In order to continue, one point needs to be clarified: the *inversion of geometry* as we cross the unit speed boundary.

As we walked to the depths of the Earth, the journey was consistently heading towards the center of a spherical shell. Half way down, gravity inverts and past the Sphere of Rest, "up" is now towards the center of the sphere, rather than outwards toward infinity, as it is on the surface.

Once we cross over the central sun boundary and enter Tartarus, our world gets yanked inside-out. We are now starting at the center of the I'M UPRIGHT "inverse sphere" in the cosmic sector, the realm of 3D time, and are AND YOU'RE going to climb up to the surface. The conventional world we left INSIDE-OUT! behind has now become buried inside the central sun of Tartarus, looking like a super-dense fluid of iron and magma, just as the core of the planet looks to everyday scientists on the Earth's surface.

Those on the surface of Earth swear they are on the outside, and Agartha is on the inside, while those in Agartha swear they are on the outside, and the Earthers are the ones on the inside.

But the reality is that *structure is an artifact of consciousness*—the way our minds interpret sensory data. In metaphysics, it is said that "everything is illusion," and that is actually quite accurate. However, if you understand the nature of the illusion being perpetrated, it becomes something real.

The environment of Tartarus is much like that of Hades, with one exception: the inner, central sun is operating from the reciprocal aspect from the other side, emitting light in the ultraviolet and X-ray bands, rather than the visible spectrum. But if you were born on that side of the boundary, it would look as a normal spectrum, since your perception would also be inverted. That's the confusing bit about insides and outsides, and when insides become outsides.

As we start to ascend to the surface of Agartha, we cross the same situations as we did on the way down from the Earth's surface. Tunneling down into Tartarus, we will reach the Sphere of Rest, based on *temporal gravity*, since we are now moving through coordinate time. Passing the Sphere of Rest, we find that gravity, once again, inverts and we are now standing with our feet towards the core and our heads towards the cosmic heavens. Again, climbing through the cave systems and eventually coming out on the surface of what appears to be another world—but it isn't, it is just the cosmic aspect of Tiamat, it's soul.

So what is different? Looking around, there is land, rivers, oceans, trees, birds, animals... golly, it looks just like the surface of Earth does. But there are a few exceptions. First, the skies are deep blue with puffy, white clouds—not a sign of chemtrails, or any other form of geoengineering. The air, itself, is clean and fresh, almost invigorating and it is quiet—only the birds and wind can be heard. No industry or technology is present, so no polluted air or energy. Life is abundant and healthy, and filled with creatures from mythology. It appears to be quite the magical realm.

After a day and night pass, you notice something missing—*no moon* in the sky. The ANNUNA have never visited this aspect of Tiamat, because their Ark, NIBIRU, also known as Earth's moon, does not have a natural soul, so it has no presence in the cosmic sector of 3D time.

A "technology" *does exist* here but it is one based on natural consequences of harmony, sympathy and discord—a type of vibratory physics that has been relegated to the realm of magic and witchcraft by the scientific minds of men. But this is the "science" here, where human, electronic technology is acting as the "black magic" of this realm.

Continuing up through the atmosphere of Agartha, we see many of the same sights. A sun in the sky, which is actually the core of Sol, inside-out, the planets and a star field. Eventually, we will encounter the outer, temporal gravitational limit at a slightly different distance from Agartha, as the Earth's moon does modify that limit with its presence here.

So ends our journey from the far reaches of space, to the far reaches of time.

### Summary of the Structure of Planets

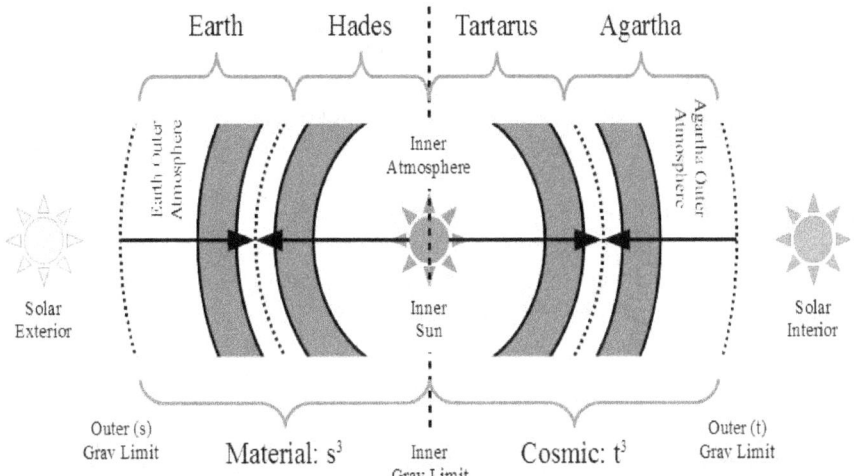

Using Earth as a model, we find that there are four realms to each living planet, two in the material half (3D space, clock time) and two in the cosmic half (3D time, clock space). The interface between these realms appears as a inner, central sun in both realms that acts more like a barrier or "guarded portal" between. Each of the realms has an atmosphere, surface and subterranean region. The Earth (body) realm has an atmosphere extending into the coordinate space around the planet, the Agarthan (soul) realm has an atmosphere extending into the coordinate time around the planet. Both have surfaces with salt water oceans and continents, covered in living organisms.

The inner realms are similar, just flipped around, with the atmospheres extending towards the inner, central sun, with a surface composed of continents and fresh water oceans, and a subterranean realm.

The inner surfaces are covered in living organisms, as well.

Credit should be given to Edmond Halley, of which *Halley's comet* is named after, for his original ideas concerning this 4-layer structure of the inner realm. His model was an attempt to reconcile the unusual properties of the moving, magnetic poles and their association to the Aurora Borealis.

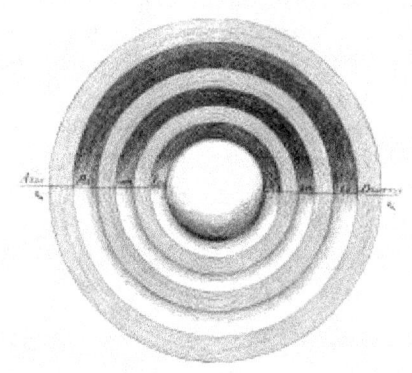

"Taking into account what had been recorded in past times, Edmond Halley discovered that the magnetic longitude variation was slowly changing. Halley's explanation of this phenomenon was to hypothesize the existence of more than one magnetic field.

"His hollow earth theory was that the earth comprised of an outer shell with a separate, inner nucleus and each of these globes had its own axis with magnetic poles and each separate axis was inclined to each other.

"The variance in the velocity of rotation of these separate globes caused the magnetized needles to seek one or another of the poles. This therefore accounted for the slow shift in the position of magnetic north. Later, when compass readings could not [be] account[ed] for by one interior earth, he proposed the world contained two more, each inside the other."[355]

Halley's model is, essentially, the same structure deduced from Reciprocal System speed ranges, where the shells form the regions of Earth's surface, Hades, Tartarus and the "inverse surface" of Agartha.

As you have seen, the model developed by the reciprocal relation of space-to-time not only addresses the conventional, hollow Earth model, but *many* of its variants. It all becomes a matter of perspective and observation, understanding the assumptions that went in to the conclusions.

### The "Flat Earth" Model

Over the last few years, there has been a resurgence in the "flat earth" theory and I will admit that I spent a few months researching the information—just in case I missed something.

When presented with an idea that opposes my world view, I *do not* consider the idea *wrong*—I consider *my* understanding of how that idea came into existence to be *incomplete*, so I do some research to understand *how* that idea was formed, by

looking at the premises that went into it. I can then compare those premises to my own and find out why the "natural consequences" took different paths. That is why I am such a big fan of the Reciprocal System, because it starts by *clearly defining the premises* on which the theory is built, as *Fundamental Postulates*.

Having done that, I can say that, yes, there is some truth to the flat earth theory, but only *some* truth. After careful consideration I found *no* evidence proving that the earth's surface, as a whole, is a flat plane but *substantial* evidence indicating that it is an oblate spheroid. However, because of the discrete unit, "quanta" of structure, it *does have flat places*, particularly in valleys, where the depression of the valley floor counters the bulge produced by the curvature of the spheroid.

Unfortunately, there is a lot of "willful ignorance" on the part of Flat Earthers that substantially detracts from any of the real arguments in support of the theory. I recently watched some videos claiming that cities could be seen 30+ miles away over water.[356] One was clearly showing Toronto, but only the upper portions of the building with all the smaller structures obscured beneath the waterline due to the curvature of the Earth. To coin a phrase, "duhhh?" It just proved the surface was curved, yet the claim was "FLAT!!!" The same thing with the recent Chicago mirage, making the same "flat" claim—yet, those that returned the next day to get photographs saw *nothing*... the city was obscured by the water again. What

---

[355] *Mystical Locations*.info, Interior World, http://mystical-locations.info/hollow_earth_halley.html

[356] https://www.youtube.com/watch?v=hlkjf07JuG4 (Seeing Toronto from 30 Miles Away – The Earth is FLAT!) https://www.youtube.com/watch?v=aLlNKy5j_O8 (Chicago Skyline seen from Michigan Proves "Flat Earth")

happened there, did the Earth decide to bend overnight? Or was it, as the TV reporters said, just an atmospheric mirage?[357] Let me "practice what I preach" and list some of the *assumptions* in the theory that led to incorrect conclusions:

1. *Light is assumed to go in a straight line.* It is well known in science that light bends in a gravitational field. Astronomers know the effect as "gravitational lensing" and it is the basis of the science of optics. Light actually slows down when traveling through a medium, such as glass and bends significantly passing the edge of an object, such as a slit. This was documented by Johann Wolfgang von Goethe in the 18th century, explaining why a spectrum only appears at the edges of light passing through a slit—not through its center.[358] If a slit in paper can bend light into a spectrum, what do you think happens to light in a gravitational field the size of a planet?

2. *The Earth is a smooth sphere.* It is a flattened sphere (technically a geoid) and not really smooth, anywhere. The gravitational pull of the Earth varies considerably across its surface, due to the different mineral content below the ground. Denser elements have a stronger pull than lighter ones. The only place where the Earth gets smooth is over the deep oceans, where the depth of the water tends to normalize out these variations (*not* on the coast).

3. *The continental crust is curved like the ocean is.* Due to the fact that the Earth is constantly expanding, valley floors tend to drop about the same amount as the curvature of the Earth rises, sometimes more, making the crustal regions a series of flat surfaces with the edges being mountain ranges, much like a polyhedron.

There are other, incorrect assumptions that go into the model, but I think reasoning might be better served if we take a look at how this "flat earth" concept arose, since it does show up in many mythological sources.

The ancient "flat earth" models have several components that make it resemble the structure of a *snow globe*, an old toy with a 3D structure contained in a transparent shell filled with water and bits of white plastic, which can be shaken to make it look like a snow storm is going on inside.

1. The Earth is a flat plane, originally expressed in ancient texts as being an island in the middle of an ocean.

2. A region of icy mountains surrounds that ocean, preventing access to the lands beyond (Antarctica).

3. The plane of Earth is enclosed in a dome, a "firmament," on which the stars are fixed. Often, an ocean is depicted above the firmament, which is acting like a glass bubble.

4. The sun and moon rotate above the plane of the Earth, like

---

[357] I grew up on the coast and have spent many, many hours on the ocean and have seen these thermal inversion mirages for myself, a number of times. And they are truly amazing to behold—but do not last very long.

[358] An excellent video on the subject is "Light Darkness & Colours – Goethe's Theory" on YouTube: https://www.youtube.com/watch?v=pitz56_8CJg

hands on a clock, being 3000-4000 miles away and ranging from 27-37 miles in diameter, depending on the researcher.

Many of the ancient civilizations consider that the Universe has this structure (Hebrew, Norse / Germanic, Biblical, Zion):

 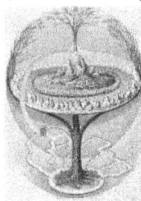  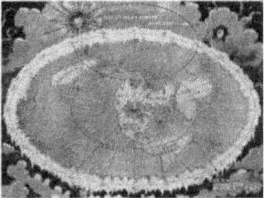

Understanding the "firmament" seems to be the key to understanding the rest of the system, for without this enclosing bubble, nothing can exist below. The common definition is:

In Biblical cosmology, the firmament is the structure above the atmosphere, conceived as a vast solid dome. According to the Genesis creation narrative, God created the firmament to separate the "waters above" the earth from the "waters below" the earth. The word is anglicized from Latin *firmamentum*, which appears in the Vulgate, a late fourth-century Latin translation of the Bible.[359]

Being a wizard, I do not rely on muggle definitions from the Wikipedia, so let's dig a little deeper into the original word, *firmamentum*: **Firmament**

from the Vulgate *firmamentum*, which is used as the translation of the Hebrew *raki'a*. This word means simply "expansion." It denotes the space or expanse like an arch appearing immediately above us. They who rendered raki'a by firmamentum regarded it as a solid body. The language of Scripture is not scientific but popular, and hence we read of the sun rising and setting, and also here the use of this particular word. It is plain that it was used to denote solidity as well as expansion. It formed a division between the waters above and the waters below (Genesis 1:7). The raki'a supported the upper reservoir (Psalms 148:4). It was the support also of the heavenly bodies (Genesis 1:14), and is spoken of as having "windows" and "doors" (Genesis 7:11; Isaiah 24:18; Malachi 3:10) through which the rain and snow might descend.[360]

Now we get to the truth of the matter, with the keyword being, "expansion." Flip back a few pages to where the *gravitational limit* was discussed, as being the *impenetrable boundary* between the progression of the natural reference system—the *expansion* of the Universe—and the local, gravitybound, 3D spatial coordinate system. "Firmament" is just the ancient, Biblical name for Larson's "gravitational limit." A firmament *does* exist, though the nature of it has been cloaked by centuries of theological interpretations. The Reciprocal System was able to lift this cloak and reveal what was beneath and it *is consistent* with the ancient descriptions of a firmament.

The big difference is that the gravitational limit is not a *dome covering a plane*, but a

---

[359] Wikipedia on "Firmament," https://en.wikipedia.org/wiki/Firmament
[360] *Bible Study Tools* on "Firmament,"
    http://www.biblestudytools.com/dictionary/firmament/

*hypersphere³⁶¹ covering a globe.* The Flat Earth model got the *concept* correct, but the *dimensions* wrong. As Emeril³⁶² does, just "kick it up a notch" and take the 2D plane of the Earth, encased by a 3D globe and move up a single dimension—a 3D sphere encased by a 4D hypersphere. Ancient texts indicate that there *is* some kind of sphere enclosing the Earth, but what about the dome shape, itself? That also seems to be ingrained into the collective unconscious and must therefore originate from somewhere important. What could be more important than the creation of man, by the Gods?

If one were to consider colonizing another planet, as the ANNUNA did here, what kind of structures would you need to get things started, where very little is known about the surrounding environment? Most human ideas start with building a dome to establish a controlled environment until sufficient research can be done to determine the external environmental properties.

*Colonizing a Distant World Mars Dome Test*

It would be logical to assume that the ANNUNA, first arriving here on Tiamat and being a lot smarter than the local chimpanzees later to become humans, would do something similar. And that is found in the Sumerian accounts of ENKI and his ABZU, the "terraforming ship" that first arrived on the planet was to set up a protected environment—a "guarded enclosure" to get basic agriculture going, later to become the center of EDIN, sort of a "Garden of Edin." And it was in this Garden that mankind was created, eventually leaving this Garden to set forth in the surrounding world—but remembering its womb-like, domed shape as his place of birth, firmly entrenched in his unconscious.

The dome surrounding the Guarded Enclosure of EDIN was designed as a barrier between the internal environment, with its local irrigation system and "grow lights" up on the ceiling to simulate a diurnal cycle, as the sun in those days was substantially dimmer and more orange, being of the K-type stellar class. Of course, the inside of the Enclosure was flat ground. So it can be seen how these two concepts became intermixed and confused.

Mankind was engineered to be slaves to the gods, but some of these household servants became knowledgeable of what their masters were doing. Some, like Adapa, were even educated by the Gods (in Adapa's case, ENKI). These human scholars were able to relate some of this information to their brethren.

---

³⁶¹ A *hypersphere* is a 4-dimensional sphere, which is the basic, quaternion rotation of [ω *ix jy kz*] with the "real" axis running between the inner and outer gravitational limits.

³⁶² Lagasse, Emeril, American celebrity chef, television personality and author, known for "New Orleans" style cooking.

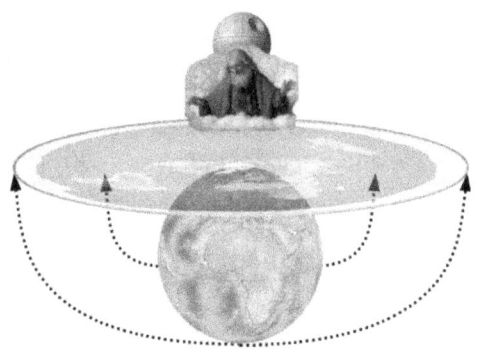

Humans tend to be curious creatures, desiring to know what is going on all around them—particularly, what boundaries hold them in place, as they were raised in such a "pen."

One of the first pieces of information to get out was a map of the world, provided by ENKI, showing the domain that the ANNUNA ruled over. They did not have holographic technology (at least for slaves), so ENKI rolled out a flat map of their domain—now called Ea's Eridu (or Earth), to show them the world. This flat map was a planar projection of the planet, as seen from NIBIRU, the ANNUNA mothership, parked in stationary orbit above the Arctic Pole. Lo and Behold, the Flat Earth model was born, circa 3500 BCE, with humanity only suspecting there was some dome shape associated with it.

Hey, this Flat Earth map came *directly* from the Gods, so it must be correct, right? Thou doth not question the Lord and live! At least that was the general consensus of the time.

In the days of the Gods, the Earth was physically smaller and the oceans were just big lakes, as can be seen in this map made by Bertius in 1618 AD. Most of the land masses were grouped together around the Bargos Islands at the North Pole, surrounded by water to the south (not shown). In the center of the Bargos Islands was a black abyss, described in the *Inventio Fortunata,*[363] where water was sucked into a maelstrom to the center of the Earth—the northern polar entrance to the Underworld.

This is why many of the ancient depictions show the "known world" as an island in the middle of a sea, because that is what it looked like on the map provided by the Gods.

At the periphery of this Flat Earth lies another impenetrable barrier of ice cliffs, standing 150 feet high and unclimbable to the old-world explorers. But modern aircraft did not have much difficulty getting over those extreme heights and mapping most of Antarctica, or as it was known in the old days, *Terra Australis Incognita*, which included modern Australia as part of the continent. These cliffs have been observed by civilians visiting the region, back when they were allowed to visit the region.

---

[363] de Linna, Nicholas, *Inventio Fortunata* (trans. "lucky discoveries"), circa 10th century. The only known copy is in the Vatican library and they aren't letting it out any time soon, as it contains too many "truths." Also described as the "northern whirlpool" in Giraldus Cambrensis' *Topographia hibernica*, circa 1200 AD.

I will be the first to admit that there are strange goings-on in Antarctica, particularly in the region of the
Hercules Dome, which sits adjacent to the southern pole and the Amundsen-Scott Base (United States). Applying some common sense, when you look at a map of the region the first thing you see is a very large number of research facilities. As of 2013, Finland, Ukraine, Argentina, United States, Uruguay, Japan, Russia, Chile, Australia, Brazil, France, Italy, India, Spain, Germany, China, United Kingdom, Poland, Pakistan, South Korea, Romania, Peru, Czech, Republic, Japan, Belarus, Belgium, South Africa, Bulgaria, New Zealand, Sweden and Norway all have research bases on this ice cube. Why? Haven't they run across snow, before? Applying some *lex parsimoniae*, somebody found something—something significant enough to make any country with the resources run down to this frozen wilderness and lay claim to some turf.

The Flat Earth theory literally "unwraps" this mystery because Antarctica is spread around the periphery of the circular plane of the Earth. The argument is made that this region is *artificial* and deliberately made inhospitable to man, so that man will not venture across the land and run into the edge of the dome covering the Earth—and the edge of the dome is what they are hiding.

The *modus operandi* of the New World Order seems to be, "make it backwards." If we've got an "upwelling" ice dome at the south pole like the Hercules Dome, and flip it backwards, you've got a hole sinking down into the Earth—an entrance to Hades and the inner realms. That may be what they are actually hiding—and very concerned about, because the population of the inner realm, like most other intelligent life, isn't too thrilled about having mankind for dinner—unless he is the "main course." The giant demons of old did like to "serve mankind," particularly with a Bearnaise.

If we take a quick look at some "safe" observational data, we know from the ancient records that the north polar opening pulled the salty, ocean water into the inner realm. Logically, the south polar opening should be the reverse—a source of fresh water from the inner realm, heading outward across Antarctica.

**Russia uncovers freshwater lake in Antarctica (Associated Press, Feb 5, 2012)**
The first indication of contact with the lake was on Saturday, but it was not until Sunday that the pressure sensors on the drill signaled it had fully penetrated the lake. Being 2.4 miles beneath the surface and 160 miles long, Lake Vostok is the largest of more than 280 known lakes in Antarctica. Lake Vostok is roughly the size of Lake Ontario.

Well, it seems we have one heck of a lot of fresh water under the ice pack, not to mention that the ice pack, itself, is *also* fresh water. The quantity of ice in the arctic is tiny, compared to that of Antarctica.

Something is definitely going on in Antarctica, but based on ancient texts and mariner records, it is more likely they are hiding a polar opening than a glass wall constructed by aliens to contain humanity. (Though, given the way most of humanity behaves, I'm sure ETs <u>have</u> considered it!)

**Laying the Flat Earth to Rest**
*If you tell a lie big enough and keep repeating it, people will eventually come to believe it. The lie can be maintained only for such time as the State can shield the people from the political, economic*

*and/or military consequences of the lie. It thus becomes vitally important for the State to use all of its powers to repress dissent, for the truth is the mortal enemy of the lie, and thus by extension, the truth is the greatest enemy of the State.*[364]

This resurgence in the Flat Earth philosophy is using this very claim, concerning the "globe earth believers," since logical arguments and observational data simply do not support the Earth being a flat plane. As usual, it is backwards. The globe model did not come from politicians, theologians or scientists, it came from *world explorers* who were actually "out there" floating around in boats for years, making maps that only worked correctly on the surface of a sphere. However, the same claim cannot be made by the Flat Earth model, which has all the "red flags" of no actual *research*, being faith-based, with the deliberate misdirection that is classic of a psy-op, a "psychological operation." I've noticed all the marketing that surrounds it and some of the pricey video productions, books and media being produced. Where there is money, there is usually a hidden agenda.

I suggest you apply some common sense and take a look for yourself. My personal opinion is that the resurgence in the flat earth theory is being promoted by *The Powers That Be*, for two reasons: to hide some "inconvenient truths" through social compliance and to see how gullible people have gotten from lack of "real" education, common sense and the almost total dependence on "socially defined" truths.

### The Nine Worlds

Now that we have a natural structure to planets, it becomes possible to map the ancient "worlds" to this structure. For this, I am going to use a bit of my own, Germanic heritage and identify the Nine Worlds of Norse mythology with modern cartography. Granted, we have been through a few expansion events[365] since then, but the basic shape and structure of the continental crust really hasn't changed all that much, outside of the courses of rivers, the occasional breaking up or shifting of a lake and a few things getting buried under ice—of which there wasn't much of in the Arctic in the 13th century, but plenty of, now (well, at least for a little while longer, as I hear the polar cap is melting away).

According to the Ásatrúar, there are nine worlds that make up the "universe." However, in ancient times, the universe was basically what mankind could see around him, and that was not a whole lot because the skies were covered in mist up until the time of the Deluge. So these "nine worlds" or *heimr*, were large tracts of land populated by a dominant species, much like *countries* of today.

---

[364] Goebbels, Paul Joseph, 1897-1945, German politician and Reich Minister of Propaganda in Nazi Germany, 1933-1945.

[365] Phoenix III, Daniel, "Geochronology: Hiding History in the Past" describes the expanding planet theory.

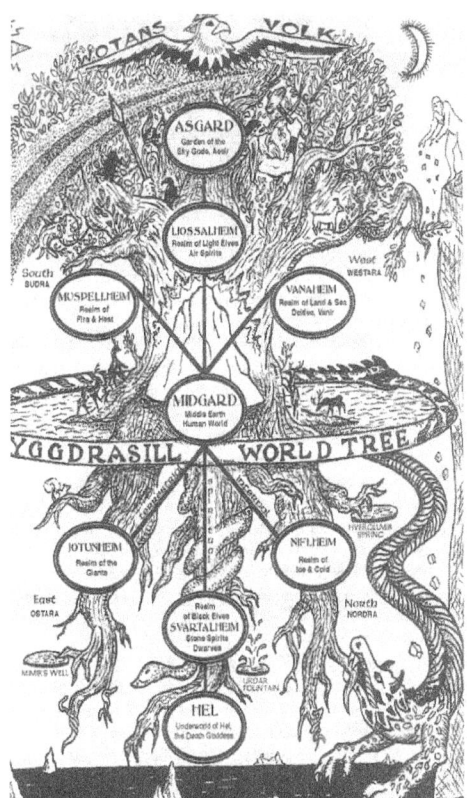

*Ásgarðr*, the home of the Æsir, is composed of two root words, Ás (God) and Garðr (garden, "guarded enclosure"). Hence its association with Heaven, the home of the gods.

*Álfheimr* or *Ljósálfheimr*, the home of the Ljósálfar, the surface dwelling L-Ms. They live in the light of day, hence the literal translation of *Ljósálfar* as, "light elves."[366]

*Múspellsheimr*, a world of fire and home of the *Eldjötnar*, the fire or "elder" giants.

*Vanaheimr*, the home of the Vanir, the Earth Gods.

*Miðgarðr*, the home of humans centered around the original landing site of the gods, ERIDU—not the entire planet.

*Jötunheimr*, the home of the *Jötunn*, the Frost Giants.

*Niflheimr*, a world of *cold mists*, mistakenly referred to as the land of ice and snow.

*Svartálfaheimr*, the home of the *Svartálfar*, the underground dwelling L-Ms that live in the darkness, the "dark elves."

*Helheimr (Hel)*,[367] the home of the "dishonorable dead," a *cold* place—not a "hot time in the old heimr, tonight."

There are also a number of other places of interest that are associated with the nine worlds:

- *Yggdrasil*, the "world ash tree," giving the relations of how these worlds connect to each other. Again, it has Miðgarðr, the "middle garden" as a flat disk with a *central mountain peak* surrounded by land with an ocean beyond, ending with *Jörmungandr*, a giant serpent that encircles the world, definining the limits of man's knowledge, as the realm as the Antarcitic ice barrier did on later flat Earth maps.

- *Niðavellir*, the lower regions of Niflheim that are inhabited by the *Dökkálfar*, the "metallurgists extraordinaire" also known as Dwarfs.

- *Bifröst*, which is commonly confused with *Bilröst* ("rainbow bridge"). Bifröst is "the shaking road to Heaven," from the old Norse *bifa*, "to shimmer or shake." Sippar Spaceport, Miðgarðr-1 requesting

---

[366] "Light" and "dark" have been misconstrued by the New Age to mean "good" and "evil," which they are *not* in mythos.

[367] Helheimr was often used instead of Hel for the realm, to avoid confusion with the ruler of the domain, also Hel.

launch clearance to Ásgarðr Station… "Yeah, take it away Ernie! Fasten your safety belts, clench your buttocks! It's going to be a bumpy ride!"[368]•
*Útgarðar*, a region of Jötunheimr popularized by the god Loki.

•        *Niflhel*, a region between Niflheimr and Helheimr.

Are these places just the imagination of drunken Bards of days of old? No, they were *real places*, once you apply a little knowledge of expanding, hollow worlds. So let's play, *Name That Lost World!*[369]

In the early days prior to the arrival of the DINGAR (the ancestors of the ANNUNA), the planet looked nothing like it does now, as there were no oceans, no moon, no cycle of day and night and was significantly smaller in size, some 4276 miles (6882 km) in diameter, about half the current size.[370] One side always faced the sun and the other stayed in darkness, but there was an equatorial region where the two blended, making a viable climate. The Norse legends begin here, with this structure.

### The Ancient Worlds

South and North were the original yin-yang of Western Europe, where north was the dark side of the hill (yin), and south was the sunlit side (yang). South is derived from the Jötunn *Surtr* (or the Hindu *Surya*), the solar guardian of Múspellsheimr. North is in reverence to *Njörðr*, the god of seafaring, as the Vikings like to travel in Njörðr's Sea (later, the North Sea), and is interesting because the "north pole" was not referred to as the *North Pole* until the 16th century. Prior to that it was the *Arctic Pole*, suggesting that there may have been a magnetic pole reversal sometime in the 13th to 14th centuries, which does occur during a planetary expansion event. These north-south "directions" occur frequently in mythology, though they tend to refer more to *orientation*, than direction.

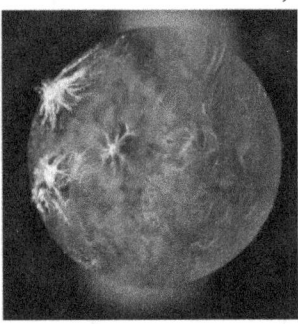

*Múspellsheimr*, the world of Fire in the south (sun-side) was the half of the planet facing the sun, being substantially larger in those days in the orange, K-type giant phase. On this side of the world, volcanoes, lava flows and hot winds abound.

On the north side of the planet, the dark half that sunlight did not reach, was a realm of ice and cold, *Niflheimr*. Moisture ejected by the volcanoes of Múspellsheimr would freeze in this arctic-like wilderness, covered in snow and glaciers.

The borderlands where north meets south and fire meets ice, was called

---

[368] Rowling, J.K., *Harry Potter and the Prisoner of Azkaban*, the Shrunken Head speaking to passengers on the shuttle.

[369] A word play on the television series, *Name That Tune* (1952-1985), where challengers tried to name a song with the fewest number of clues.

[370] Because the earth expands and forms oceans, the land area remains about the same, being 75,470,000 mi². Using that as the surface area of an ocean-less sphere, the radius comes out to 2138 miles, or a diameter of 4276 miles. This is approximate, as the current ocean level has submerged coastal regions of continents (the continental shelves), which are not included in the figures for land area.

*Ginnungagap*, the first place of habitation in Norse mythology, a temperate zone running across the terminator between light and dark:

*Ginnungagap, the Yawning Void… which faced toward the northern quarter, became filled with heaviness, and masses of ice and rime, and from within, drizzling rain and gusts; but the southern part of the Yawning Void was lighted by those sparks and glowing masses which flew out of Múspellsheimr.*[371]

After man was created, he learned of these stories from the gods, but lacked the understanding to put them in their correct places in time and space. So, he applied the *same names* to places that he knew about, with similar properties. The reuse of names is very common in mythology, so one must look at the context behind the label to determine what is appropriate.

There is not much to reveal about these ancient worlds, as they were boring places. The building blocks of life were being forged in fire and ice, preparing this small planet for the evolution of consciousness. Our world was much like a fertile valley, waiting for the explosion of life to begin. Of course, fertile farmlands tend to attract farmers, and in our case, these farmers were some monks from a solar system, close, close nearby.

### Welcome Stranger

*Last week as you recall, Will Robinson had sent out a radio signal, unaware that far out in the void of space, a strange, missile-like object was even now homing in on it…*[372]

Tiamat, the 3rd rock from a young, reddish-orange star recuperating from a recent supernova, had been colonized by a race remembered as the *Cyclopeans*, an extraterrestrial species of giant hominids that were one of the very first species to evolve a high state of consciousness and move out amongst the stars. The Cyclopeans had advanced to the point where simple reincarnation could no longer find a suitable expression in their structure of body, mind and spirit, and the Cyclopeans had colonized Tiamat as a monastic sanctuary to research the concept of ascension for their species—but not the typical "ascension of density" as described in metaphysical research, but ascension totally out of our octave of existence, into a new realm, going beyond space and time as we now understand it.

The process of this spiritual research had affected the local environment causing a burst of life to occur on the world in its earliest stages along the tropical zone of Ginnungagap. Out of this burst of life evolved several intelligent species known to "insiders" as the L-Ms, the "Little Men," and to folklore as the mythological creatures of old: the faeries, dwarfs, elves, trolls, goblins and their ilk, along with other larger hominids, the great apes (with and without tails), the latter developing into Neanderthals.

The Cyclopeans, being on a humble and spiritual path researching that "meaning of life" stuff, were never treated as gods by the L-Ms. Instead, they were simply guides and advisors, assisting the development of new consciousness on this world where they could, working towards the goal of *living in rapport* with one's

---

[371] Sturluson, Snorri, *The Prose Edda*, translated by Arthur Gilchrist Brodeur, 1916, p. 17

[372] Turfeld, Dick, narration to the 1965 *Lost in Space* episode, "Welcome Stranger."

surroundings. This became part of the evolutionary path of the L-Ms, their *modus operandi*, forming more of a "brotherhood" psychology than one based on domination and submission. As a natural consequence of this relation, they never developed the concept of *worship*, and no energy was put into "placating the gods" to keep from being destroyed. The entire, collective intelligence of this L-M brotherhood was directed towards the advancement of consciousness and the evolution of their species—and they evolved, very quickly—learning how to coexist with their somewhat hostile surroundings and limited space along this belt of life between the fires of Múspellsheimr and Niflheimr.

Over the course of centuries, the planet went through a number of expansion events, the sun got smaller and warmer, moving towards the orange-yellow spectral classes, and the extremes of fire and ice began to subside, creating more habitable zones in those formerly off-limits realms. As a consequence of this, the L-Ms evolved along species lines that paralleled the "elements" of evolution: *fire, water, air* and *earth*. These "elemental" species gave rise to a number of races, each particularly suited to their local environments, and in harmony with them.

However, nearing the end of their million-year quest into the mysteries of ascension to the *Theta Octave*[373] of existence, another less-evolved race of space-faring extraterrestrials dropped by for a visit, hoping to claim the resources and riches of this new world for their own—an amphibious race of genetically altering shape-shifters known as the S-Ms, the "Space Men" or "Saurian Men," to the ancient Greeks as the *Titans* or the Sumerian, DINGAR.

DIN.GAR

A small expedition from the DINGAR Ark, called NIBIRU, was sent to this 3rd rock from the sun curious as to why there *was* a significant amount of life in this otherwise rocky world of fire and ice. The evolution of this world, which they referred to as KI,[374] appeared to be out of sync with the rest of the solar system. And when they got here, they discovered why… an ancient race of giants "beat them to it," having already established colonies on KI.

And the DINGAR were totally puzzled as to why the Cyclopeans weren't stripping the world for its resources and riches—as they had intended. Really, why would anyone want to bother with this ridiculous "spiritual" stuff, when there are fields of diamonds and crystals, lining the banks of flowing rivers of gold and precious metals? These Cyclopeans must be crazy!

---

[373] According to the surviving Archive of the Cyclopean race, our civilization exists in the *Eta Octave*, the seventh evolutionary stage of consciousness, which is one of the many reasons that "7" is prominent in metaphysics.

[374] Pronounced "key," as in Qi or ch'i, meaning "life."

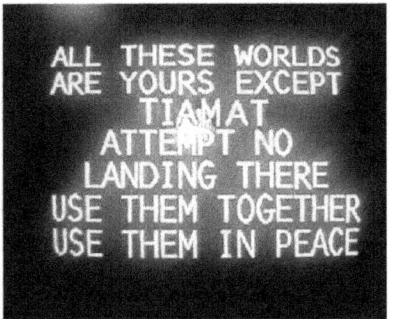

ALL THESE WORLDS ARE YOURS EXCEPT TIAMAT ATTEMPT NO LANDING THERE USE THEM TOGETHER USE THEM IN PEACE

After some negotiation, a deal was cut between the Cyclopean/L-M population of Tiamat and the DINGAR, letting them do whatever they want with the other planets and asteroids in this new solar system, as long as they kept clear of Tiamat, as to not interfere with the advancement of consciousness on this world. The DINGAR, realizing that there was "no way in Múspellsheimr" that they could defeat the Cyclopeans, accepted the compromise, packed up their shuttlecraft and returned to NIBIRU, setting it on a course for the 4th rock from the sun, the planet we now refer to as Mars.

The Cyclopeans and L-Ms were left to continue their studies in peace and rapport, evolving in body, mind and spirit. After some years, the Cyclopeans finally uncovered the path to Ascension out of the Eta Octave, and departed, leaving this world to those that evolved from it. A few of the Cyclopeans did stay behind to keep an eye on things, to make sure the DINGAR kept their word about the quarantine state of Tiamat. But, like most politicians, the word of these "soon-to-be-gods" meant very little.

### The Terrestrial Worlds of the Gods

With most of the Cyclopeans gone, the DINGAR set about their plans to exploit the resources of KI for themselves. AN,[375] the youngest of the DINGAR and eager to establish his position in their hierarchy, was chosen to lead an expedition to KI to establish a base of operations and start agricultural exports. AN's symbol is the *sickle*, the reaper that prepares the harvest, and was known for his tiara of horns. From these, we get the imagery of the Grim Reaper and the crowns of Kings.

With two, new Arks being constructed in Martian orbit (used to stabilize the environment), AN and his sons took NIBIRU across the firmament to KI, parking it in a stationary orbit above the northern pole of Niflheim, at the L2 Lagrangian point. He then sent his eldest son, EN, to KI, to begin colonization.[376]

---

[375] Later known as the Greek Cronus, the Roman Saturn, the Norse "All Father," Borr, the Vedic Shiva, the Mayan Ahau, the Christian, "God the Father (of Jehovah)," and many other names.

[376] EN is translated as "Lord," but a more accurate translation is "#1 Son," (to borrow from Charlie Chan) whom was normally lord and master of a new household. When EN took up residence on KI, he became ENKI, the first born of AN on KI, or later, the Lord of Earth.

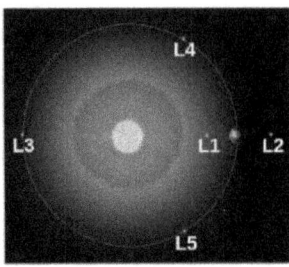

Parking the Ark in this L2 location was no accident, it was part of a well thought-out plan on the part of AN. They knew that as young planets *cool*, they *expand*, due to the supernova fragments forming their cores that were moving in time, rather than space. In those days, KI was only about twice the size of NIBIRU, a doubling of diameter, making KI an "octave" higher than NIBIRU. The Ark was positioned to use this resonant structure and timed to match an expansion event on KI.

Expansion events disrupt all the static conditions of a world, flipping magnetic poles, altering positions of land masses and even starting, stopping or changing the direction of rotation, just as a natural consequence to balance the new energetic arrangement. In the case of KI, a non-rotating, fire/ice world, the imbalance of such a large, celestial body as NIBIRU caused a magnetohydrodynamic effect to take place and start the planet spinning on its axis. But the rotational axis was not on the line from L1 to L2, but orthogonal to its orbital plane, alternately exposing the frozen waste of Niflheimr to the sun and Múspellsheimr to the cold and darkness.

Within a short period of time, the hot regions cooled down and the cold regions warmed up. This caused flooding of the surface from melting ice, spread around the world filling in natural basins, creating many lakes and rivers and nourishing the volcanic soil. Life was taking full advantage of it, spreading wildly across the globe, evolving and adapting to the new environments.[377]

This shaking up of the world shook up a lot of things, knocking both the remaining Cyclopeans and the L-Ms off balance. But it paved the way to colonization by AN and his NUNA, his children, referred to as the ANNUNA.

The DINGAR, as a species, had a lot of experience with colonization and terraforming new worlds, as they have done it many times in the past. So they were ready to proceed with colonizing KI as soon as the opportunity arose. EN took his terraforming ship, ABZU ("creation wisdom") and established a "guarded enclosure" in one of the newer regions of KI, away from the local, intelligent life. This place became AN's "home away from home," ERIDU.[378] With this background, we can now determine how the ANNUNAKI, the "ANNUNA living on KI," spread across the still expanding globe, divvying up the lands and responsibility amongst the Sons of AN.

One of the major side-effects of the rotation of Tiamat was the creation of strong belts of magnetism, surrounding the world—much like spinning the armature of

---

[377] This was recorded in the geologic record as the *Cambrian Explosion*, purportedly 542 million years ago. Correcting for "hiding history in the past," this was fairly recent, in the vicinity of 250,000 years ago. The Sumerian Kings List documents the reign of the ANNUNAKI as being some 241,200 years in length, prior to the Deluge.

[378] After ERI.DU was abandoned by the ANNUNA, it was left to the children of the gods, the BAI, becoming the "children's home," DU.BAI. The original colony is *still* Dubai, in the United Arab Emirates (UAE).

a generator. These belts, which we refer to as the Van Allen belts of radiation, trapped many of the ionizing particles from the sun producing some very dangerous travel regions for a space-faring civilization. To expedite travel to and from NIBIRU, the Ark was moved to a new location over the rotational pole where the intermediate-speed magnetic lines of force were their weakest. Less magnetic concentration meant less damage to their shuttlecraft when passing through it.

Realizing that mythology (and religion) are describing the invasion and colonization of our world by a hostile species—a very common motif in Science Fiction stories—we can correlate the ancient worlds to modern understanding.

The mysterious worlds of the ancients aren't so mysterious, once you put them into context.

### Ásgarðr, the Garden of the Gods

In the Norse tradition, the warring sky-gods were the Æsir, whom resided in Ásgarðr, up in the heavens. The Æsir (ANNUNA) are the Sons of Borr (AN), that live in a self-contained realm of

Ásgarðr (NIBIRU), the "guarded enclosure of the gods."

We have already discovered that these "guarded enclosures," *Garðr*, are typically spherical or hemispherical shells protecting an environment. On a large scale for space travel, these are known as "Arks," a hollow, supernova remnant that has been converted into a biosphere for travel between stars and planets. Ásgarðr is one of these

Arks that is *still in equatorial orbit* around Earth—our Moon.

(Why *Death Star* do you think there was such a "space race" to get there—it's full of *Moon* the technology of the gods!)

There are many "dead Arks" floating about our solar system, from millennia past. After all, the ANNUNA did have the run of the solar system for quite some time.

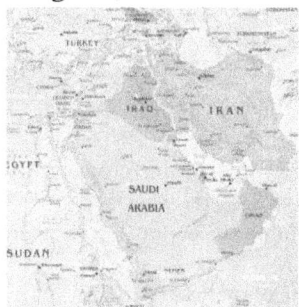

George Lucas picked up on the basic "Ark" design in *Star Wars* as the *Death Star*. In

the Moon photo, the circular *Mare Imbrium*, the Sea of Showers, was edited with the *Iapetus* Death Star's primary weapon reflector, just to highlight the similarity. Notice a *pattern* here?

### Miðgarðr, the Middle Garden

Miðgarðr was the first ANNUNAKI *country*, so to speak, growing up around

ERIDU near the original "guarded enclosure" of

EDIN.[379] Referred to as the "middle garden" because it was in the middle of the new, habitable zones, and later on was considered the half-way point between the gods above in Ásgarðr and those below, in Hel (Hades). The center of Miðgarðr is the modern city of Dubai,[74] the original location of ERIDU, with cultural development spreading along the river valley to the modern countries of Kuwait and Iraq, all the way to Syria and Cyprus as a Mediterranean port.

There were a number of major city-centers in the region, known as *böll* (halla) or "great halls." These are documented extensively in the Sumerian records, unlike the more poetic Norse tales:

- ERIDU, the first city and operational headquarters.
- BADTIBIRA, the mining facility.
- LARAK, the Space Flight control center, much like NASA at Houston.
- SIPPAR, the ancient version of Cape Canaveral.
- SHURUPPAK, the hospital complex.
- URUK, built specifically for the DINGAR, so AN and his siblings had somewhere nice to stay.
- NIPPUR, the administrative center of government.

The region of the Middle East (the Miðeast, so to speak) looked very different in those days, being a fertile land of plenty—not rocks and desert. Today's Persian Gulf was nothing more than a river, with the mouth near Muscat in Oman.

### *Bifröst, the Shaking Path connecting Miðgarðr with Ásgarðr*

This photo has a shape similar to a rainbow, doesn't it? But quite a "bumpy ride!" because it is a photo of a *rocket launch*, not a bridge made of light.

Any Sci-Fi/Fantasy geek worth his wand will have figured out what Bifröst *is*, from my earlier comment—the shaking path that rockets and shuttlecraft follow, moving supplies and personnel between Miðgarðr colony and the orbital Ásgarðr "mother Ark." The illuminated trail of fire and smoke arcing across the sky is the bridge.

In Sumerian, this launch pad was known as SIP.PAR, the "bird city," and was the "stairway to the stars" run by SHAMASH,[380] the Chief Astronaut, ruled by King

---

[379] EDIN means "watered plain," most likely referring to the hydroponic crops grown under the dome. The ANNUNA were original vegetarians. After work began outside of the dome, EDIN took on the additional meaning of the fertile valley in which ERIDU was constructed.

[380] Also known as Utu (Akkadian and Semitic). After the Deluge, BAALBEK became the new launch site.

Enmeduranna,[381] "the Lord whose MEs[382] bond Heaven and Earth."

On the other side of Bifröst is the Ásgarðr hanger deck, under the control of *Heimdallr,* who was considered a bit of a "hunk" by human standards, since most of the gods looked like horny toads. Heimdallr's position as Hanger Deck Chief gave him a lot of contact with human slaves and he took an interest in their plight, helping to define human social classes—not really as an act of kindness, but of *efficiency,* but it is easier to offload cargo when there is a clearly defined hierarchy of responsibility.

Note that there is nothing *spiritual* here. Though Bifröst *is* the path connecting Earth to Heaven, it is just a bumpy rocket ride on a cargo transport or shuttlecraft. It has nothing to do with the "ghost road" or the transport of the souls of the dead into heaven. Fruits, grains, cattle … they were transported across Bifröst. When a human took that ride, it was usually because they were in a *lot* of trouble and were being taken to appear before AN for judgment. Don't believe me, ask ENKI's servant, Adapa. ENKI taught him the "forbidden knowledge" of the gods and when the "south wind," the ANNUNAKI version of a helicopter, buzzed him while he was out fishing and flooded his boat, he used that knowledge to break the wing of the wind, crashing it into the sea. That earned him a ride across Bifröst! AN was really pissed that a mere human, "broke wind" (so to speak).

These two worlds formed the initial colony on Tiamat now being referred to as EA's ERIDU, later shortened to EARTH and its linguistic variants. But what happened to the existing, intelligent life that was already there, along with the remaining Cyclopeans, the *Eldjötnar?*

Consider what the Europeans did to the indigenous tribes of the Americas: either be *enslaved* as the African population was, or be *exterminated* as they attempted to do to the Native Americans. There are many stories of the attempted enslavement of the Mayan and Aztec people of Central and South America—but they would not have it. Horses were not native to the Americas, so the Spanish tried to ride the Mayans like animals. The Mayans response was interesting… "sure, hop on" and then they would jump off the nearest cliff, taking their rider with them. The response of the Spaniards was to gift them lots of blankets and other material— taken from leper colonies.

Agriculture was booming for the ANNUNAKI, using the slaves they brought with them, the IGIGI, a race from the *azonei,* a region outside of DINGAR control. However, they were running into problems, namely the operation was getting too big for the IGIGI to handle, so some of the less important ANNUNA were conscripted to help out. This led to a lot of internal conflict within the gods. Attempts were made to enslave the L-Ms, as they were a bit naïve about the ways

---

[381] Enmeduranna is most likely the Biblical Enoch, as the stories concerning them are very similar.

[382] A "ME" (pronounced "may") is normally translated as a "tablet of destiny," which the gods often fought over and stole from each other. There were many of them. To possess a ME was to possess a specific power over what that ME controlled. Think of it more like a tablet PC with an encrypted controller App that only can run on a specific tablet.

of the universe and could easily be scammed into providing valuable work for the colonists—and they were highly skilled with the resources of this world, having the "magical" knowledge of the Cyclopeans. But, they were treated just as the Native Americans that survived the conquest of America were: confined to their own world, Álfheimr, the "land of elves," conveniently placed just below Ásgarðr, parked in orbit directly overhead. After all, the gods needed to keep an eye on these rebels, as the remaining Cyclopeans did encourage them not to do as the gods demanded.

The Cyclopeans continued to cause trouble and the increasingly impudent ANNUNAKI were getting rather upset that they were doing all the work, while the ANNUNA sat around on the Mother Ark, drinking beer and watching sports. AN realized he was losing control of the situation, and sent word back to his DINGAR brothers on Mars, requesting assistance.

The DINGAR took one of the new Arks across the firmament to Earth, giving the planet two moons for a while. Negotiations took place and it became apparent that AN had gotten a little over-ambitious, operating for himself, outside the plans of the DINGAR. The colonization was not supposed to have upset the native life—and AN went and literally turned the world upside-down. Concern arose that the ascended Cyclopeans might return and boot them all out, so the DINGAR made peace with the resident Cyclopeans and told AN to "take a hike." AN did not take this sitting down, gathering his troops and he made a preemptive strike against his brothers and the Cyclopeans—and won.[383]

### The New Múspellsheimr, Tartarus

With mining operations well under way in Helheimr (Hades), the gods knew of the barrier across the internal sun, the realm of Tartarus. There was but a single crossing, an anomaly that occurs where coordinate space and coordinate time coexist in a region that allows passage. And that is where they sent the captives, the surviving DINGAR (the Titans) and the Cyclopeans (Eldjötnar, Cyclopes), to imprison them for eternity.

This region became the new Múspellsheimr, the realm behind the inner sun, blocked by an impenetrable, natural barrier with the single access point well protected.

The inner sun is basically part of the projection of the "inside-out" core of the external sun, which at this time, was young, orange and just getting started on its evolutionary journey. All the planets have these inner sun cores, as does every atom of matter (the tiny emissions there are radiation). The inner sun, itself, was more of a dull, reddish glow, and the many, mineral-rich magma flows made both Helheimr, and Tartarus on the flip side, look like Hell.

### The New Niflheimr (Mist World) and Niflhel (Misty Hell)

Niflheimr is the region surrounding the Sphere of Rest, that existed much as it does today—a region of a strange, atomic mist, with two possible directions of *ascent*, either down to Helheimr (Hades) or up through the mantle to the surface. Two of these three regions are part of the Norse worlds:

- *Svartálfaheimr*, the region we call the "mantle" between Niflheimr

---

[383] Described by the gods after the creation of man as the *Titanomachy*.

and the surface. This is the realm of dwarfs, Nature's miners.

- *Niflheimr*, the misty region surrounding the Sphere of Rest.
- *Niflhel*, not normally a "world;" the stony, cave region between Niflheimr and Helheimr.

### Jötunheimr, The Home of the Nephilim

The Æsir, the Sumerian ANNUNA, where known to have had "genetic relations" with other species on the planet, mixing DNA cocktails to see if they could get a workforce to replace the gods and IGIGI whom were becoming increasingly discontent. In later years, ADAM, mankind, replaced those workers and the gods were quite pleased with themselves and continued their genetic relations with mankind, producing a race of giants—part human, part god—called the NEPHILIM.

These giants ended up breeding like rabbits and were put on a reservation, far away from the L-M and human populations now covering the northern and equatorial regions of the planet. This place was the "Land of the Giants,"[384] *Jötunheimr*, at the southernmost regions of the world, a place marked on the ancient maps as *Terra Australis Incognita*, the "unknown lands to the south," which we now know as Antarctica and Australia, a single land mass in those days on a much smaller planet.

Much of Jötunheimr was a rocky, forested wilderness without much animal life. The NEPHILIM lived primarily on fish, caught in the abundant, freshwater rivers that flowed out of *Mímir's Well*, the reciprocal of *Rupes Nigra et altissima* in the Arctic—the back door to Hel, currently located under the Hercules Dome in Antarctica.

The peninsula that formed today's Australia was the capital of Jötunheimr, named *Útgarðar*, ruled by Skrýmir. Now, if this assumption is correct, Útgarðr would have the highest population of giants, so there should be some giant remains in Australia, today.

### Fossil Australians could have been 12 ft tall and 600 pounds[385]

*In old Pleistocene river gravels near Bathurst, New South Wales, huge stone artifacts— clubs, pounders, adzes, chisels, knives and hand-axes-all of tremendous weight, lie scattered over a wide area. A fossicker searching the Winburndale River north of Bathurst discovered a large quartzitised fossil human molar tooth, far too big for any normal modern human. A similar molar of chert fossilisation was also recovered from ancient deposits near Dubbo, N.S.W. Prospectors working in the Bathurst district over 40 years ago frequently reported coming across large human footprints in shoals of red jasper.*

As it turns out, giant skeletal remains, tools and structures are scattered all over

---

[384] Allen, Irwin, *Land of the Giants*, TV series, 1968-1970. A human spaceship crashes on an Earth-like planet, to discover the inhabitants are twelve times their size.

[385] Gilroy, Rex, "And There Were Giants," *Psychic Australian*, October, 1976.

Australia, making it an accessible part of *The Land of the Giants*. The ancient mariners never ventured into *Terra Australis* because of the giant population, so much of the interior went unmapped. There are a few surviving records, most of which came indirectly via India and China—ports of call for the merchant vessels of the Jötnar. What these maps show is a C-shaped region with a large lake surrounding the South Pole, from which water poured forth from the depths of the Earth.

### Vanaheimr, The Realm of the Vanir

The Vanir, the "earth gods," are the Sumerian ANNUNAKI, the Sons of AN that stayed on KI, the Earth, led by ENKI. In Old Norse, ENKI is called *Vili* and his younger brother, ENLIL was Oðinn. In the Sagas, Oðinn was considered "first born," but if you compare to the Sumerian records you find that ENKI, Vili, was actually first born and when he rebelled against his father, Borr, was "demoted" to #2 son moving ENLIL, Oðinn, to #1 son. Ranking was very important to the gods, as it is in any military organization. And the confusion does not end there, as Vili *Vanaheimr and Aztlán* and Vé (the third son of Borr), never carried much influence in mythology
and were later incorporated into the gods Þórr (Thórr or Thor) and Freyr, retaining the "big three" (formerly: Oðinn, Vili, Vé, later: Oðinn, Þórr, Freyr).

Knowing this, we find that Thórr was the "protector of mankind," as was ENKI, whom was the one that told Noah (Njörðr) to build the Ark—not the Hebrew God, Jehovah, who is ENLIL. ENLIL *wanted* mankind destroyed, along with the NEPHILIM. But that is another story…

Vanaheimr was the home of the gods that decided they *liked* living on Earth and abandoned their celestial heritage to stay here. This caused a split resulting in the Æsir "sky gods," the ANNUNA, and the Vanir "earth gods," the ANNUNAKI. Dear Ole' dad, in order to prevent global conflict, split up the domains of the warring brothers to opposite sides of the world. Oðinn got Miðgarðr, the Middle East, and Vili/Thórr got Vanaheimr, the *Americas*.[386] Vé ended up with region around China, just to keep him clear of the battle.

Since ENKI was the "Chief Science Officer" of the gods, most of the technological advancement occurred in Vanaheimr, the Americas. Being a very

---

[386] ENKI wasn't the best looking god, being remembered as "the hideous one" and was amphibian in stature, very serpentlike. As a consequence, his land was referred to as "the land of the serpent," *Amaruca*, later anglicized to *America*. If you do some minor digging in old records, you find that Amerigo Vespucci was born *Alberigo* Vespucci, later changing his name to Amerigo after the discovery of Amaruca—to keep people from knowing its true origins. America wasnt named after Vespucci—Vespucci was named after America! See, everything you know is backwards.

fertile country, farmers (the Native Americans) were created to handle agriculture for the gods. After orichalcum[387] was discovered in the southern regions,[388] Olmec miners were brought from Africa (the original mining site) to begin mining in South Vanaheimr. This put a LOT of power and raw materials under ENKI's control—and he decided to rebel against his father, AN, and brother ENLIL, and take over the planet for himself.

ENKI, known as "the accuser" (*Satan*) or the "enlightened" (*Lucifer*), befriended the L-Ms of the northern and coastal regions, to create an army against ENLIL. His headquarters, Aztlán (Atlantis) was *not* in Mexico, but on the eastern coast of the northern region of Vanaheimr, now known as: *Washington, DC*, the center of Lucifer's empire in the Land of the Serpent. (I'm sure that comes as no surprise.) If one searches for information on Vanaheimr, one will find *very little* information, as all of this has been purged from the history books. The correlation between Vanaheimr, the empire of the serpent gods, and Amaruca, the Land of the Flying Serpents, was not easy to find—but traces do remain.

So, we have the planet divvied-up now, regions ruled by specific gods, with one Lost World left to identify, the one that the gods did not have much control over.

### Álfheimr: The L-M Reservation

I've saved the best for last. Of all the Norse worlds, Álfheimr is by far the most intriguing because it is one, huge contradiction. Digging through the Norse Sagas, there is very little information on Álfheimr, other than it is the home of the Álfar, the "Elves," or as referred to in these papers, L-Ms. Yet, millions of fables concerning the "Little People" exist all across the Americas, Europe and Asia. So why does Álfheimr remain such a information-loaded mystery? Well… it seems that a lot of effort was put into concealing it. So let's pull off this "cloak of invisibility"[389] and see what is hidden beneath.

First, here are some of the physical characteristics of Álfheimr that we can use to locate that lost world:

- It is the world nearest to Ásgarðr, the Lunar Ark over *Polvs Arcticvs*.
- It was divided into four lands of different races of Álfar.
- It was adjacent to Vanaheimr (the Americas) and Miðgarðr (the Mideast) was to the south— remembering that "south" is "sunnier."
- Álfheimr was divided into two worlds, Ljósálfheimr and Svartálfaheimr, a world above in the sunlight and a world below, in the

---

[387] Orichalcum is reddish, radioactive gold—nuclear fuel—which was what the ANNUNA were mining until the Earth's magnetic ionization level dropped and gold became a relatively stable element, with uranium taking its place. See: *Geochronology: Hiding History in the Past*, page 14, available on http://conscioushugs.com

[388] The original mining operation was in the mountains east of Lago Poopó in Bolivia, from where we get a lot of the "Atlantis" (Aztlán) mythos.

[389] From the *Harry Potter* series; Harry had a cloak that would render its contents invisible.

darkness. In conventional interpretation, Ljósálfheimr is considered synonymous with Álfheimr, but that is like referring to "space" as "motion," which leads to the confusion of "how can I have motion without any**thing** moving?" The time aspect of motion went unconsidered, just as the Svartálfaheimr realm of Álfheimr goes unconsidered.

- Svartálfaheimr was adjacent to Niflheimr.

What we already know is that Ásgarðr was in a fixed location above the Arctic (North) Pole. The closest point on the planet to that location would be the region *surrounding* the Arctic Pole. There would be four "landts" in this region being adjacent to Vanaheimr (*Amaruca*) with Miðgarðr, the Middle East, being on the opposing side, further to the south.

With that in mind, we can identify Álfheimr on the old maps: referred to in 17ᵗʰ century Europe as the ***Bargos Islands***. I believe the name actually refers to the *Barge Landt*, the "dangerous shores," as the islands were surrounded by sharp, mountain peaks and the four rivers were treacherous to sail with swift currents, often dragging ships to their doom. "Barge" is not in reference to a flatbottomed boat, but to "barge in," to bump into or collide with, in a rude or clumsy way.

The Bargos Islands are four islands that exactly surround the Arctic Pole that appear on all maps prior to the early 17ᵗʰ century, all over the world. And these were maps used by seamen to navigate the oceans. The lands are described as fertile and green, surrounded by snow-covered mountains near the shorelines and the local inhabitants were "pygmies," little people that stood about a meter tall.

One of the curious features about the "Barge Lands" is in the very center of the map stands a mysterious, black mountain, *Rupes nigra et altissima*, the "cliffs black and deep." The renowned Wizard of the 16ᵗʰ century, John Dee,[86] noticed this on Mercator's 1604 map and wrote to him, asking for a description of his comments, which were from *Inventio Fortuna*.[59] Mercator replied:

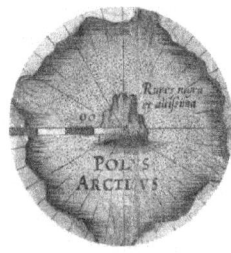

> *"In the midst of the four countries is a Whirl-pool, into which there empty these four indrawing Seas which divide the North. And the water rushes round and descends into the Earth just as if one were pouring it through a filter funnel. It is four degrees wide on every side of the Pole, that is to say eight degrees altogether. Except that right under the Pole there lies a bare Rock in the midst of the Sea. Its circumference is almost 33 French miles, and it is all of magnetic Stone… This is word for word everything that I copied out of this author years ago."*

Not only does this "Island at the Top of the World"[87] occur in many cultures, so does this strange mountain and abyss, the *Maelstrom*. In Vedic lore, it is *Mount Meru*, or *Sumeru* in Sanskrit. In the Norse, it is the trunk of the world tree, Yggdrasil.[88] There was (or is) something there at one time, guarding the entrance to the inner worlds.

So, the big question becomes, "where are the Bargos Islands, now?" since there is nothing but deep, cold water at the rotational North Pole.

The answer: the Earth *expanded*, splitting open above Canada and Russia, creating the Arctic ocean and sliding the Bargos Islands southward towards the Atlantic in compensation. The expansion changed its climate drastically and the islands started to freeze over, as the logs of Leif Erikson indicate—the Viking colonies established there moved to the coastal regions, as the fertile land in the interior became covered in snow. The Bargos Islands, Álfheimr, is STILL HERE, not only buried in snow, but buried in a lot of "bull."[89]

What was done to conceal it was to "stretch" the extend of a surviving island to the south, *Groenlandt* (the "green land"), to cover up the glacial cover-up,

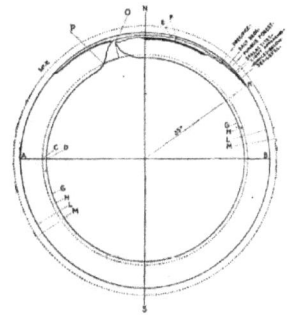 absorbing the Bargos Island as the northern parts of *Greenland*.

I located a 1981 National Geographic map of what Greenland looked like, under the ice pack. The resemblance is striking—even the original rivers are there. And the Rupes Nigra is also still there, though the shifting land mass has probably blocked off all access to where the water once went. This is also shown on the map in *Etidorhpa*,[43] how the entrance to the inner world has slipped away from the North Pole (label "O"), now sitting under a glacier in northern Greenland.

86   Dee, John, born July 13, 1527, died 1608. Advisor to Queen Elizabeth I, and the guy that created the agenda for the New World Order.

87   Marshall, James Vance, *Island at the Top of the World*, screenplay by John Whedon.

88   See the 2nd picture from the left on page 22, showing Yddrasil as the world "axis."

89   "Bull" is an American Old West reference, referring to cattle manure that you don't want to step in, walking down the street.

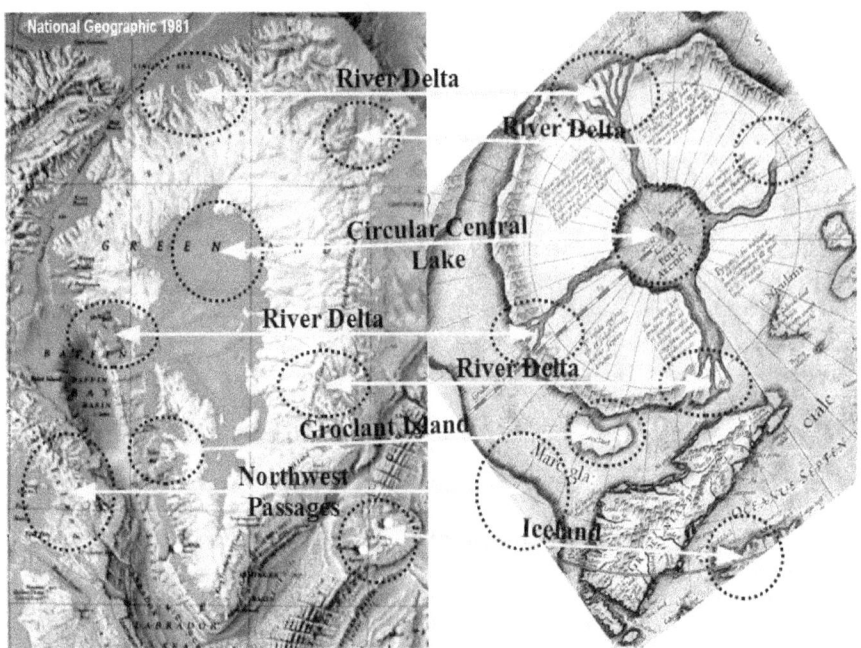

This correlation has been missed by many because of the overuse of the flat, Mercator map projection, which greatly "stretches" the lands in the polar regions, distorting their shapes.[390]

As we've postulated that the Earth underwent an expansion event, using these maps we can find out just how much the Earth expanded because the maps of the 16th and early 17th centuries have the locations of large cities that are still in the same place today. We need to identify two locations that are on the same land mass, as expansion tends to expand ocean and sea beds—not the continental crust. That means the distance, today, should be about the same as it was prior to the expansion.

Let's pick a couple of cities near the perimeter of the map "Polvs Arcticvs siue Tract, Septentrionalis," such as the west coastal city of Bergen, Norway and the east coastal city of Stockholm, Sweden, as they are near the perimeter next to the longitude lines and on opposite coasts of the same land mass. The map shows Bergen at 26 degrees and Stockholm at 42 degrees, approximately.

Today's distance between Stockholm and Bergen is 446 miles, line of sight. The edge of this map is at 60° north latitude, so the circumference around the Earth at that point would be smaller than the equatorial distance, approximately 10,035

---

[390] I happened to have a world globe when I was a kid, so I noticed, many years ago that Greenland did not look like it did on flat maps. To me, it looked like a big question mark.

miles:

446miles

×360°=10035miles

16°

As we are looking at a flat map, the center of that circle at 60° would be where the rotational axis is, circumference = 2π radius, or, radius = circumference / 2π:

10035miles

=1597miles 2 π

That would make the equatorial radius:

1597 miles

=3194 miles cos(60)

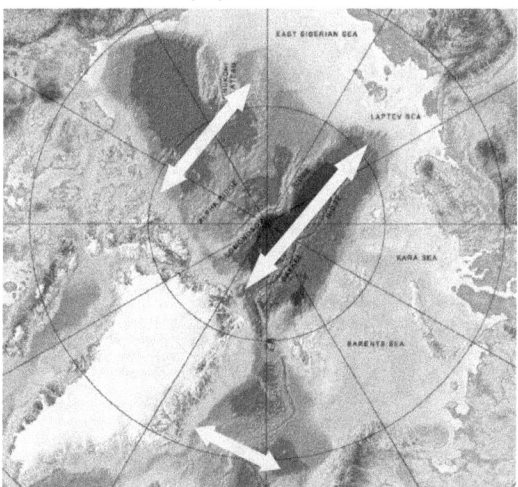

The current, equatorial radius of Earth is 3,959 miles. That means this map was made for a planet about 81% of the current size of Earth, only 20,070 miles around the equator, some 5,000 miles smaller in circumference than it is now. We know that continents tend to crack, but overall, don't stretch as much as the ocean floor does, so that means that there is about 5,000 miles of deep ocean that cracked open somewhere. Notice these nice, deep, fresh cracks in the Arctic ocean floor, with the nice, smooth bottoms?

I am proceeding with the assumption that the land area remains constant, simply because of the way the continents plug together like pieces of a puzzle, and it is only the ocean floor that is expanding.

The Arctic Ocean is all deep water and accounts for about half of those miles, indicating that the Earth expanded in other locations, too, down along the mid-Atlantic rift, widening the north Atlantic, putting more distance between the original Groenland and Ireland and the U.K., sinking a few islands along the way, such as the old-World trading center, *Frislandt*.[91]

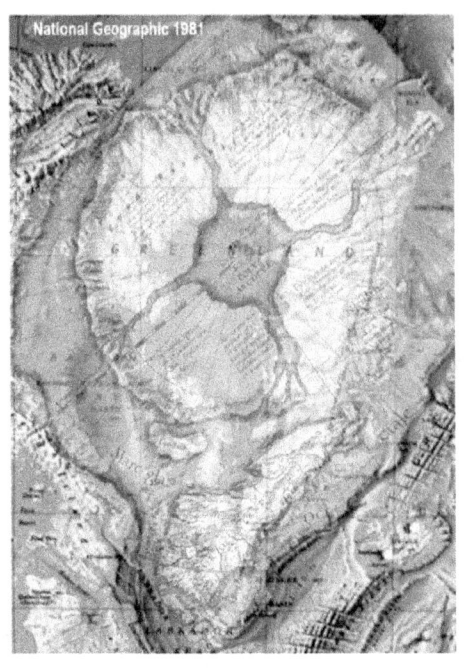

To verify, a rough calculation of land area outlining the Bargos Islands, including Groclandt and Groenlandt, comes out to approximately 830,000 square miles. Greenland, today, is 836,300 square miles. Mystery solved! With that known, we can now identify who's who and what's what on the old 1604 map. The land left of Groenlandt across *Frecum Danis* is Newfoundland. *Mare Glaciale* is Baffin Bay.

Schetlac Island would be the Faroe Islands, Scotia is Scotland, Eislandt is Iceland, and off to the left, *Hic Mare est dulcium aquarium* is Hudson Bay. Many familiar features are included on this map, so why would mariners invent "phantom islands" like Frislandt? *Lex parsimoniae*, they did not. These are real places and since Stockholm is ON the map and was founded circa 1250 AD... the Earth hasn't been the size it is for very long, again, hiding history in the past.

Additional information was found in a rather curious people, the Irish Tuatha Dé Danann, whom brought four, magical gifts with

91 Frislandt is yet another "phantom island" that was sunk during the expansion event, as many of the smaller, volcanic islands were, such as the neighboring islands of Neome and Fodalida. The island is now the sea mount west of Ireland, located at approximately 56.8 N, 17.5 W. If you examine ocean floor topography, you can see the rip right between Groenlandt and Frislandt.

them to Ireland: *Dagda's Cauldron*, the *Spear of Lugh*, the *Lia Fáil* (Stone of Destiny) and the *Claíomh Solais* (Sword of Light), from four "cities": *Findias (or Finias), Falias, Gorias* and *Murias*.

*The Tuatha Dé Danann had traveled to the "Northern Isles" where they learned many skills and magic in its four cities Falias, Gorias, Murias and Findias. From there they traveled to Ireland bringing with them a treasure from each city.*

—Lebor Gabála Érenn

I've quoted "cities" because the references are from the 11th century, and the idea of a "city," a cathedral town, did not really come into use until the 13th century. In the older days, it would be more of a community or commonwealth, or any place where such a community formed—such as an island.

The Tuatha Dé Danann are an interesting people, as their home was in the *Otherworld*,[92] which we have identified as the Greek *Hades*, the interior of the Earth. It would make sense that the Tuatha Dé Danann would arrive through the Arctic polar opening, the *Rupes Nigra*, and end up in Álfheimr, the Bargos Islands with the L-M Álfar.

By examining various Irish tales concerning the journeys of the Tuatha Dé Danann, the "northern islands" parallel the four lands and races of Álfheimr, which is quite uncanny, particularly since these same types of magical gifts were the specialty of the L-Ms that lived there. So, as we've been playing "Name That Lost World," we can also do "Name That Phantom Island."

Correlation was done based on *common attributes*, not directions, because the Earth has undergone a few changes since then and may not have been rotating in the direction it is today. This map is prior to at least one expansion event, maybe two. In the early 1600s, the island listed as Murias had virtually disappeared under water and the original maps had to have the lower right part of

**Falias**

that island erased and modified to account for the few, surviving islands that mariners were now finding. Yes, Murias became a sunken city, just as the folklore tells.

*In the frost-grown city of Falias lit by the falling stars*
*I have seen the ravens flying like banners of old wars—*
*I have seen the snow-white ravens amid the ice-green spires*
*Seeking the long-lost havens of all old lost desires.*[93]

The magnetic pole is listed on two places on these old maps, one as a mountain and the other nearby as a location. With this structure, the Earth, being smaller and with a weaker magnetic field, would have the Aurora Borealis *directly over* Falias, the "lit by falling stars" and "ice green spires."

92 In Celtic mythology, the *Otherworld* is the realm of the dead. This is not the same as the *Other Realm*, the Cosmic sector of 3D time.

This island was the home of the Ljósálfar, the "light elves," because the sky was never dark. They were[391] experts in the domain of "the fire of life," able to heal and cure, and had the gift of immortality. Many of the stories of a "fountain of youth" on a remote island originate here.

---

[391] Macleod, Fiona, "The Dirge of the Four Cities," *Poems and Dramas*, Vol. VII.

The gift of Falias, the *Lia Fáil*, was called the *Stone of Destiny*—or in Sumerian, the "Tablet of Destiny," the ME.[78] This magical item was said to have two powers: first, when a rightful King's feet were placed upon it, it would "roar in joy." Second was the power to rejuvenate the king, so that he may have a long reign. These ME were tuned to a specific genetic marker, so only the direct descendants of the ANNUNA could use them. "Rightful kings" were directly descended from the gods, with these markers.[392] The roaring of joy was the activation of the device, given the proper genetic key.

### Gorias

*In Gorias are gems,*
*And pale gold,*
*Shining diadems*
*Gathered of old*

The treasure of Gorias was the *Spear of Lugh*, later called *Lúin of Celtchar*, a radiant energy weapon along the same design as *Mjölnir*, Thórr's hammer, built by the Dwarfs of the region as this marshland was the primary entrance to Svartálfaheimr, the underground realm of the Dwarfs. As the mineral merchant capital of the world, gems, gold and other precious minerals were plentiful, custom crafted into diadems and scepters by the dwarfs for the overlords.

### Findias

*In the torch-lit city of Findias that flames on the brow of the South*
*The Spear that divideth the heart is held in a brazen mouth*

This was the most fertile and luscious of the four islands, a literal Faeryland, the land of the Álfar and a major trading port for Europe and Amaruca.

The gift of this land was the *Claíomh Solais*, the "Sword of Light" on which the lightsaber of Star Wars fame is modeled. This was not actually a weapon, but a masonry tool known as a *force cutter*, that could cut any material by separating its atoms like a knife through butter.[393] However, it did not take humanity long to figure out that it could also cut the toughest armor in half, without any effort. This sword and the Lia Fáil later ended up in Europe, being "the sword in the stone," *Excalibur*.

---

[392] This genetic marker key was played upon in the *Stargate SG-1* and *Stargate Atlantis* series, where only those with the "ancient gene" could activate Ancient technology.

[393] Such a device was reproduced in the late 19th century by researcher John Worrell Keely.

**Murias**

*In the sunken city of Murias A golden Image dwells:*
*The sea-song of the trampling waves*
*Is as muffled bells Where He dwells.*

Murias was originally the "city of eternal sunset," inhabited by the Nøkk, the Álfar of the deep that sided with ENKI in his war against ENLIL, and are often seen depicted on statuary with Poseidon (ENKI). The dwellers of Murias were normally a peaceful people possessing great skills in music and harmonics.

The Nøkk had many treasures of the deep ocean, including a type of yeast which was used to create Cornucopia, "Horns of Plenty," which the Tuatha Dé Danann took as the *Cauldron of Dagda.* These bottomless containers of food are Elven bread, a highly nutritious yeast that when supplied with any type of sugary liquid (mead) and sunlight would grow at capricious rates, producing mushroom-like bread in considerable quantity. Life finds a way.

### *Yes, Virginia, there IS a Santa Claus… And he lives in Lemuria.*

When reading stories of the sunken city of Murias, one cannot help but notice the similarities between the Irish "Na Murias" and "Le Muria," also a sunken continent. Could it be that Lemuria is actually Murias in Álfheimr, the partially sunken Bargos Island, just north of Europe?

Let's look at some history on Lemuria, which is a reference to the French *la mer,* "the sea," anglicized with the "-ian" suffix to infer "the people of the sea," known in Norse as the Nøkk—water sprites. Northern Europe is full of legends of water sprites, the neck, nicor, nixie, nokken, nikker, nekker, näck, näkki and many other names. And it is primarily *northern* Europe, the lands that would be adjacent to the Island of Murias. So it is likely that the "people of the sea" were the people of "na Murias," carried across many different dialects to end up as the modern Lemurians. Even the old translation of the abbreviation "L-M" meant "Lemurian-Muanian," the people of Lemuria and Mu. (Mu is the "motherland," which we know is Miðgarðr, the Middle East—not the middle of the Pacific.)

The confusion on the location of Lemuria/Murias almost seems deliberate, similar to the concealment of information regarding Álfheimr and the "phantom islands" in the Arctic. Both polar regions seem to be "off limits" to anyone except those approved by The Powers That Be.

What appears to have happened is this: in 1864, zoologist Philip Sclater wondered why lemurs could be found in both India and Madagascar, but not in adjacent Africa. He hypothesized that India and Madagascar were at one time a single continent, which he named *Lemuria,* after the lemurs. He was actually right about the continent, which is documented in ancient Indian records as *Taprobane.* So the original name of the sunken continent, Na Murias, was overwritten by Sclater's Le Murias as Lemuria.

Now that we know that "Lemuria" was in the *Arctic* (not the Indian ocean) and populated by elvish water sprites that brings up another interesting European faery tale, that of a bunch of elves living at the North Pole that would bring food and gifts to the children of Europe during the winter.

Yes, Virginia, there IS a Santa Claus and he lives in Lemuria near the North Pole. Ho Ho Ho!

## *Epilogue*

*History is the lie commonly agreed upon.*

—Voltaire

The more one digs into the depths of religion and mythology, the more one is forced to a few, simple conclusions:

1.      Religion has nothing to do with spirituality, it is just an historical account of the colonization of the world. The Church and State were intentionally created to control your Soul and Body. True spirituality is an aspect of consciousness that "the powers that be" still cannot reach.

2.      Mythology, including "phantom islands," "lost worlds," and "pseudo-science" are actual, truthful accounts of what has been going on in the past—buried in misdirection and *fantasia*, to keep people ignorant and having to rely on "experts" to do any thinking for them.

3.      Our mental world is *not* an illusion, as Guru's like to say, but a *deliberate misdirection*. People realize this unconsciously, which is why films like *The Matrix* are so popular. Yet, in response, humanity just pulls out their box of blue pills and starts munching away.

The information that I have presented in this paper comes from many, many sources, tied together by the basic, reciprocal relation between space and time that forms the core of the *Reciprocal System of theory*. And I do want to point out, since it has been repeatedly asked concerning my prior papers, that NONE of this information is "channeled" or comes from any source outside of our own world, nor does it come from any "Extra-dimensional Entities from the 24½th Density!"[394] For the most part, it's just from old books laying around in libraries. Granted, there is a lot more intelligence on this world than most people realize— if you know where to find it. And I've left a trail of clues in this paper on doing exactly that.

It seems that the collective effort of humanity is to *hide the truth*—not *discover* it— and the human race now lives in the State of Denial, a very overpopulated location. Examine the way people think these days—it is just regurgitation from someone else, which is regurgitation from someone else, which is regurgitation from someone else, *ad nauseum*. Imagination and creativity, which are the real passions of life, are gone. People have been trained *not* to have opinions—only be good slaves and promote the opinions of their masters. What we've ended up with is what Derren Brown refers to as *social compliance*, "truth by popular consent," a total disconnect from any Natural consequences. Since humanity is abandoning Nature, it won't be long before Nature abandons humanity. And for most people these days, their world will come to an end when they can no longer get 5 bars on their smartPhone.

---

[394] A spoof on the cartoon, "Duck Dodgers in the 24½th Century," which is a spoof on the TV series, "Buck Rogers in the 25th Century."

And you are probably thinking, "something needs to be done about this!" And you are correct. But it will not be done by "The Powers That Be," the New World Order, Religions, New Age or any person of notoriety. The only way new information will ever be discovered is if people finally decide to stop accepting everything they have been told and work together to discover the truth for themselves. And it is up to us "peasants" to do it.

The Hierarchy is designed to stop all inconvenient truths that might disrupt their power and profits, so a reasonable approach would be to set up a situation that is not based in power or profit, but concern and compassion for the personal advancement of humanity. A kind of Shangri-la or Monastic Sanctuary, based in spirituality and the evolution of consciousness, with the goal of figuring out "what IS mankind's potential?" Are we meant to be slaves for eternity, or is there something more?

What do you think?

I read an interesting paper last night written by Miles Mathis, titled, "What I Finally Understood."[395] The gentleman has it right and can be summarized in just one line from his paper:

*This is what I finally understood: all famous people are there to misdirect you. ALL OF THEM.*

I, for one, am tired of living in this world of deception and being led around by the nose by famous experts and want to discover the truth behind our existence.

---

[395] Mathis, Miles, "What I Finally Understood." http://mileswmathis.com/guru.pdf

www.ingramcontent.com/pod-product-compliance
Lightning Source LLC
Chambersburg PA
CBHW080908170526

45158CB00008B/2039